The Chemical Treatment

of Boiler Water

James W. McCoy

Supervisor, Refinery Services
Standard Oil Company of California

Chemical Publishing Co.
New York, N.Y.

ISBN 0-8206-0377-5

Chemical Publishing Co., Inc.

2nd printing, 1984

Printed in the United States of America

Preface

This book, which has been in preparation for nearly 30 years, is an attempt to present in an organized and useful way the principles of an important branch of industrial chemistry. In this discussion of the chemistry of boiler water treatment, I have been as quantitative in the presentation as possible. There is, of course, no shortage of published material in this field using the descriptive approach, but I believe that chemists and chemical engineers prefer a quantitative treatment, when such is possible. I am sustained in this belief by numerous favorable comments from readers of my previous publications. Accordingly, the chemistry of boiler water treatment is offered here in terms of solubility products, formation constants, depletion rates, partition coefficients, and dissociation constants. A set of numerical problems is also appended with which the reader can test his comprehension of the text; the solution of these problems should not be difficult for anyone who has a fundamental understanding of boiler water treatment.

As the chemical aspects of treating feed water and the mechanical features of boilers and auxiliary equipment are inextricably linked, I have provided an introductory chapter in which the operation of a steam generator is briefly explained. Also included are discussions of the regulations that govern the operation of pressure vessels, the properties of steam, the composition of boiler feed water, and the more important relationships among operating variables such as steam production, blowdown, and make up rates. It seems to be the custom of boiler operators to speak of steam pressures in terms of psi when their meaning is gauge pressure, i.e., psig; I have followed this custom unless absolute pressure is meant, in which case psia is used.

Next, a chapter is included in which the objectives in treating feed water and boiler water are defined. Among these are the prevention of scaling, corrosion, and the contamination of produced steam. In the ensuing three chapters details of the numerous procedures for achieving these objectives are considered. These are classified, for convenience of presentation, as physical, external, and internal methods of treatment. The latter, which are applied to the water within the boiler, include clarification, prevention of foam, and control of scaling and corrosion within the boiler.

Chapter 6 contains brief descriptions of several types of boilers and an extensive discussion of the more widely used programs of water treatment. Throughout, I have confined the discussion of these programs to combinations of specific chemicals, eschewing the proprietary approach in which, more often than not, purportedly new methods of treatment turn out to be new methods of marketing. Then, in Chapter 7, I have chosen to discuss a number of complications that arise in the operation of boilers, among them furnace corrosion and deposits, internal scaling and deposits, superheater and turbine fouling, preboiler corrosion, tube failures, and contamination of boiler feed water. The final chapter contains control procedures for use in boiler plants, together with a number of analytical methods for analyzing chemical cleaning solutions, evaluating ion exchange resins, and assaying formulations of neutralizing amines. A glossary is also appended.

In this book I have given specific operating conditions and procedures for treatments with which I have had personal experience. In general, these are my own views of how these programs should be managed, and no other individual or organization is responsible for them, nor indeed, constrained to agree with them.

I wish to express my appreciation to my associate Mr. Roger P. Lescohier, of the Chevron Shipping Company, for many instructive discussions dealing with the operation and water treatment of marine boilers. My wife, Dolores, has as usual been of great assistance during the final preparation of this work.

Richmond, California   James W. McCoy
January 2, 1981

Contents

Chapter 1.

Principles of Steam Generation

The treatment of water, for whatever purpose, is a relatively abstract subject unless considered in terms of the end use of the treated water. Thus, before any practical understanding of the chemistry of boiler water treatment can be acquired, it is first necessary to have some knowledge of how a steam generator and its accessories function. In general, the ensuing discussion is confined to direct-fired, naturally recirculating, bent-water-tube, steel boilers, although a few other types are mentioned. For a comprehensive treatise on the mechanical aspects of boilers and associated equipment, the reader is referred to the work of Shields.[1]

Technically, a boiler is a device, consisting of a containing vessel and heating surfaces, for transforming water to steam. A source of heat—typically, a furnace—is also required, as well as some auxiliary equipment including, among other things, fuel burners, feed water pumps, draft fans, and controls. In operation, a steam generator converts the potential energy of fuel into heat energy, which is then transferred to steam for some specific purpose. Steam is used for heating, stripping and other industrial processes, and to drive turbines, pumps, compressors, and reciprocating engines. Steam for industrial processes is usually delivered at 150 psi or less, but the economical generation of electricity requires a minimum pressure of 600 psi. Some utility boilers have been erected that operate at 5000 psi, but because of metallurgical considerations, most current designs are for 3500 psi maximum. The production of process steam and electricity in the same plant has lately come to be called co-generation, particularly when waste heat is used to generate the electricity.

Fig. 1.1 is a schematic diagram of a two-drum, bent-tube boiler that is used to illustrate the operation of a steam generator. As can be seen, the unit consists of two drums, one above the other, connected by tubes. The surfaces of those tubes exposed to the fires are referred to as the primary or radiant surfaces; temperatures here are 2200–3000 F. Water-wall tubes

1

are also provided to absorb radiant heat—up to 450,000 Btu/h/ft² —thus cooling the refractory walls that enclose the furnace. Combustion gases at 500–1500 F are conducted through flues or ducts to tubes further back in the furnace, where additional heat is transferred by convection to secondary heating surfaces. In addition, carbon dioxide and water in the flue gases radiate a certain amount of energy directly to the tubes in the convection section.[2] Radiation refers to the transmission of heat in the absence of a material carrier; convection describes the transmission of heat by currents within a fluid, either a gas or a liquid.

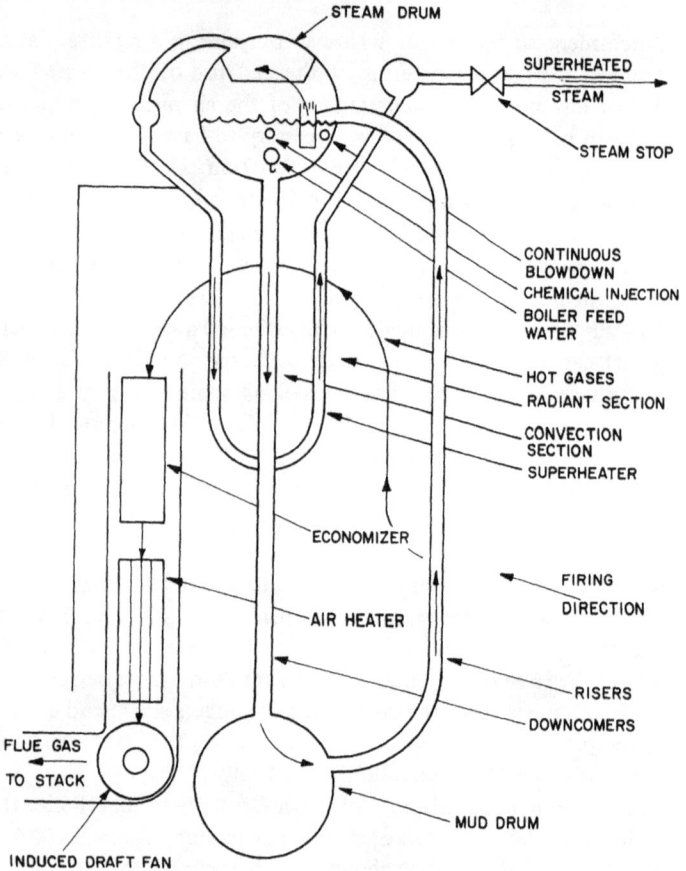

Fig. 1.1 Diagram of a 2-drum, bent-tube boiler

Although equipment for the recovery of waste heat is not always provided, the efficiency of a boiler is considerably increased if the air used for combustion and the feed water to the boiler are preheated by exhaust gas from the furnace. In the arrangement shown in Fig. 1.1, feed water is heated by flue gas at 300-800 F in an economizer, then the gas at 300-600 F heats air for combustion in an air heater, after which the cooled gas is blown to the stack by an induced draft fan.

Fig. 1.1 shows that the steam drum, downcomer tubes, mud drum, and riser tubes comprise a closed hydraulic loop. Suppose now that this assembly is filled with water and a flame is directed against the risers. As the mud drum and downcomer tubes are situated so that they are not exposed to the fire in the furnace, the density of the water in the risers decreases relative to that in the remainder of the loop. The denser, cooler water in the downcomers, therefore, pushes the warmer water up the risers to the steam drum, thus initiating a natural circulation. Eventually, as more heat is applied, steaming commences, and when the boiler reaches its operating pressure, steam, after further heating in the superheater tubes, is ready for use. The purpose of the superheater is to transfer additional heat to the steam and to evaporate the last traces of water.

As water in a boiler evaporates, salts in the remaining liquid become more concentrated, requiring that some of it be drawn off to prevent foaming, precipitation of salts, and other difficulties. Chemicals are added to help control these undesirable developments, and feed water is supplied continuously to replace losses by evaporation and blowdown. All of these processes take place in the steam drum.

1.1 REGULATIONS GOVERNING THE OPERATION OF BOILERS

The operation of pressure vessels is regulated by various state, county, and municipal laws, as well as by insurance and other codes. The American Society of Mechanical Engineers has formulated a code for operating boilers[3] of such exemplary standards that it has been incorporated into the laws of many states. As the primary objective of these rules is to ensure that pressure vessels are operated safely, their enforcement usually falls within the jurisdiction of a state's Department of Industrial Safety. The rules are formulated by the ASME Boiler and Pressure Vessel Committee and are enforced by inspectors commissioned by the National Board of

Boiler and Pressure Vessel Inspectors; the National Bureau of Casualty Underwriters looks after the interests of companies that insure boilers.

By law, a boiler must be inspected both externally and internally once a year, the "Permit to Operate" expiring one year from the date of the internal insepction. If, however, near the anniversary of the internal inspection, an inspector finds upon external examination that circumstances warrant, he may issue a temporary permit (sometimes called a variance) to operate the boiler for an additional six-month period, following which a second extension may be granted. To qualify for these variances, the operator must demonstrate that he has a superior preventive maintenance program. Also, the steam generator must be under the supervision of a competent water treatment specialist who maintains proper levels of treating chemicals and keeps records of the tests and methods used to maintain the concentrations of chemicals in the boiler water within acceptable ranges. These records must be accessible to the boiler inspector at the time the variance is requested. The purpose of all this is to ensure that pressure vessels are operated safely, so, obviously, chemical treatments and operating procedures must be followed that are acceptable under the provisions of the Boiler Code and that suggest to the inspector that the operator of the boiler has some idea of what he is supposed to be doing.

1.2 PROPERTIES OF STEAM

At every pressure up to the critical pressure there is a corresponding boiling temperature of water. As steam at that temperature and pressure is saturated with heat, it is called saturated steam. Precisely speaking, saturated steam contains no moisture; thus, if heat were added, its temperature would be raised. Steam leaving the steam drum always contains some water, however, so steam quality is defined as the ratio of the weight of vapor to the total weight of the steam-water mixture. Steam containing 3 percent of water is referred to as commercially dry; its quality is 97 percent. Purity is measured by the concentration of solids carried in the steam. If this concentration exceeds a few hundredths of a part per million, salts may deposit in superheater tubes and turbines, leading to tube failures and impaired efficiency.

Heating steam at a given pressure to a temperature higher than the saturation temperature corresponding to that pressure (as given in steam tables[4]), is called superheating; the incremental rise in temperature is the degree of superheat. There are two reasons for superheating steam: 1) The process removes water droplets, which at high velocity erode and pit turbine blades; 2) The efficiency of a turbine increases as the difference between the throttle temperature and the exhaust temperature of the steam increases. Thus, superheated steam does useful work through the entire range of superheat without condensing as it passes through successive stages of a turbine.

Table 1.1 contains values of temperature and incremental heat realized when one pound of water is heated from 32 F at atmospheric pressure to 598 F at 400 psi. Enthalpy is the number of British thermal units required to raise the temperature of water initially at 32 F and atmospheric pressure to some other specified temperature and pressure.

TABLE 1.1

Enthalpy-Temperature Relationships

Step	Temperature (F)	Incremental heat (Btu)	Enthalpy (Btu)	State	Pressure (psi)
—	32	—	—	liquid	0
1	212	180	180	liquid	0
2	212	970	1150	gas	0
3	448	54	1204	gas	400
4	598	100	1304	gas	400

The sequence of events in Table 1.1 is as follows:

Step 1. One pound of water at 32 F is heated to its boiling temperature, 212 F, by adding 180 Btu.

Step 2. An additional 970 Btu of heat are added, which initiates boiling. This is the latent heat of vaporization.

Step 3. The vapor being produced after *Step* 2 is contained, and the addition of an additional 54 Btu raises the pressure of the steam to 400 psi and its temperature to 448 F, the saturation temperature.

Step 4. The further addition of 100 Btu at 400 psi superheats the steam to 598 F.

As a consequence of these properties of steam, it costs only 10 percent more for fuel to operate a boiler at 1250 psi than it does at 400 psi.[5]

In doing work, steam expands, decreases in pressure, and cools, then, when it reaches its saturation temperature, it begins to condense. To avoid wet steam in a turbine, the partially expanded steam may be returned to the furnace for reheating before passing through the final stages of the turbine. Most boilers have only one reheating cycle, but those operating above the critical pressure, 3203.6 psi, may have two. The reheated steam, of course, is returned to the turbine at whatever pressure it had reached when extracted, but with more heat added. A supercritical boiler described as 5000 psi, 1200/1050/1025 F delivers steam to the throttle of a turbine at 5000 psi and 1200 F; after the first reheat cycle the temperature is 1050 F (at a pressure around 1025 psi); and after the second reheat the temperature is 1025 F (at a pressure of perhaps 235 psi).

Because steam does work as it expands, it is advantageous to expand it as much as possible; that is, to exhaust it from a turbine at the lowest possible pressure. A condensing turbine exhausting into a surface condenser, for instance, produces a substantially larger volume of steam than a turbine exhausting to atmospheric or higher, pressure. Steam entering a surface condenser is condensed to liquid by cooling water, producing a vacuum. Five hundred cubic feet of steam at 35 psi and 280 F, for example, expands to 1600 ft^3 at atmospheric pressure and 212 F, but to 4600 ft^3 when exhausted into a vacuum of 20 in. of mercury in a surface condenser.

1.3   WATER FOR GENERATING STEAM

Having very briefly outlined the function of a boiler in generating steam and also a few properties of the steam itself, we now turn to the subject of the book. The quality and characteristics of the water fed to a steam generator to replace that lost by evaporation and blowdown have a profound effect on the performance, life, and safety of the generating equipment. Noll[6] has summarized the chemical and physical developments within a boiler that are influenced by the composition of the boiler water. These include corrosion, carryover of solids and water into the steam, the volatility of silica, the solubility of turbine deposits, the formation of scale on heat transfer surfaces, the adherence of sludge to the walls of tubes and headers, and the buffering capacity of the boiler, which is its ability to resist the effect of contaminants on the alkalinity of the water.

In general, feed water for a boiler is made up of two components: returned condensate from steam generated in the boiler itself, and additional water that has been treated to remove or reduce the concentration of undesirable chemical ingredients: the second is called make up. Central power stations have a very high rate of condensate return (perhaps 99 percent) consequently they use little make up, whereas industrial steam plants that supply process steam may require nearly 100 percent of make up.

In the next few chapters various processes for conditioning water for boilers are described in some detail. Many schemes have been proposed for treating water, most of them accompanied by case histories (invariably eminently successful), testimonials, and promises of trouble-free operations. It should be recognized, however, that the success of any program of water treatment depends upon a fundamental understanding on the part of the boiler operator of the process of steam generation and scrupulous control of water quality. Without these prerequisites, no chemical method of treating water can forestall serious trouble, regardless of assurances to the contrary. If, through ignorance, indifference, or inattention, some particular treatment is improperly applied, there is no reason to suppose that some other program will give any better results. More often than not, complications that arise in the operation of boilers stem from shortcomings in application rather than in the method of water treatment.

1.4 RELATIONSHIPS AMONG OPERATING VARIABLES

A great deal of guessing is done in the average steam plant about variables that, with a little effort, can be exactly calculated. To illustrate, suppose that the boiler depicted in Fig. 1.1 is operating normally, that is to say, the boiler is full of water, steam is being produced by evaporating some of the boiler water, and feed water is being added continuously to replace that evaporated. It can readily be seen that as this process continues, salts in the feed water increase in concentration in the boiler water. Because of considerations explained in Chapter 2, a certain maximum value is selected for the concentration of total dissolved solids in the boiler water. When this concentration has been reached, a valve in a drain from the steam drum is opened, allowing a certain volume of boiler water to bleed off continuously to a flash tank, hot well, or atmospheric drain. By comparing the concentration of total solids in the feed water with that in the boiler water, we can tell how much concentration has occurred; the concentration of total solids in the boiler water divided by that in the feed water is called the blowdown ratio. Sufficient feed water must be supplied to the boiler to replace that lost as blowdown and evaporated as steam.

In order to establish some quantitative relationships that exist among the variables in the operation of a boiler, the following symbols are assigned:

R = the blowdown ratio
V = the volume of water in the boiler (gal)
f = the feed water rate (gal/h)
s = the steaming rate (gal/h)
b = the blowdown rate (gal/h)
$(TS)_f$ = total solids in feed water (ppm)
$(TS)_b$ = total solids in blowdown water (ppm)

It is apparent that

$$f = s + b \qquad (1\text{-}1)$$

and also, that when the boiler reaches equilibrium at a preselected value of $(TS)_b$, with b adjusted to hold that concentration, then the weight of dissolved solids entering in the feed water equals the weight of dissolved solids removed in the blowdown water. Thus,

$$f(TS)_f = b(TS)_b \qquad (1\text{-}2)$$

and

$$f/b = (TS)_b/(TS)_f = R \qquad (1\text{-}3)$$

Dividing Eq. (1-1) by b gives

$$f/b = s/b + 1 \qquad (1\text{-}4)$$

Thus

$$R = s/b + 1 \qquad (1\text{-}5)$$

and

$$s = b(R - 1) \qquad (1\text{-}6)$$

If now a certain weight of some chemical not present in the feed water, e.g., disodium phosphate, is added to the boiler, and then the concentration of phosphate in the boiler water is determined, the total capacity, V, of the boiler can be calculated. If the weight of the chemical added is W

pounds, its concentration, c, is W/V pounds per gallon. Suppose 7.5 lb of Na_2HPO_4 is added to the boiler, then after a few minutes another sample of boiler water is taken, and the concentration of phosphate, PO_4^{-3}, is found to be 24 ppm. Since disodium phosphate contains 66.9 percent of phosphate, the pounds of phosphate added is 5.0. Therefore,

$$5.0 \times 10^6/V \times 8.33 = 24$$

or

$$V = 25,000 \text{ gal}$$

It is found by chemical analysis that as time passes, the concentration of phosphate, c, decreases continuously. If the rate of depletion of phosphate can be exactly calculated, disodium phosphate can be added at the same rate, or at suitable intervals, to maintain its concentration at some desired value, or more realistically, within a specified range of values.

To derive an equation for the rate of depletion of chemical, the following must be noted:

1. The steaming rate, s, has no effect on c, because as water evaporates it is continuously replaced by an equal volume of feed water.
2. The weight of chemical in the system is unaffected by adding feed water, for the latter contains none of the treating chemical.
3. Chemical is removed in the blowdown water at the rate of bW/V lb/h.

As W/V is the concentration, c, the rate of change of c with respect to time, t, can be expressed by a differential equation, the minus sign indicating depletion.

$$dc/dt = -bc/V \qquad (1\text{-}7)$$

Separating variables and integrating,

$$\int_{c_o}^{c} dc/c = -b/V \int_{t_o}^{t} dt \qquad (1\text{-}8)$$

or,

$$\log_e c \Big]_{c_o}^{c} = -(b/V)t \Big]_{t_o}^{t} \qquad (1\text{-}9)$$

Thus,

$$\log_e (c/c_o) = -b(t - t_o)/V \qquad (1\text{-}10)$$

If, after 3 h have elapsed, the concentration of phosphate is again determined and found to be, e.g., 21 ppm, b can be calculated from Eq. (1-10).

$$\log_e (21/24) = -b \times 3/25{,}000$$

$$b = 1100 \text{ gph}$$

Suppose next the concentration of silica in the feed water and in the boiler water is determined and found to be 1.4 and 28 ppm, respectively. The blowdown ratio is then

$$R = 28/1.4 = 20$$

and from Eq. (1-6) the steaming rate is calculated to be

$$s = 1100(20 - 1)$$

$$= 20{,}900 \text{ gph } (174{,}000 \text{ lb/h})$$

Also, from Eq. (1-1), the feed water rate is

$$f = 20{,}900 + 1100$$

$$= 22{,}000 \text{ gph}$$

By making four simple chemical determinations R, V, f, s, and b can be evaluated, a knowledge of which, as is shown in Chapter 6, enables a boiler operator to maintain very close control of a steam generating plant.

REFERENCES

(1)   Shields, C. D. 1961. *Boilers: types, characteristics, and functions.* New York: McGraw-Hill.
(2)   McAdams, W. H. 1954. *Heat transmission.* New York: McGraw-Hill.
(3)   American Society of Mechanical Engineers. 1965. *Boiler and Pressure Vessel Code*, Section I, *Power Boilers*, and Section VII, *Suggested Rules for Care of Power Boilers.* New York.
(4)   Keenan, J. H., Keyes, F. G., Hill, P. G., and Moore, J. G. 1969. *Steam tables: thermodynamic properties of water including vapor, liquid, and solid phases.* New York: John Wiley & Sons.
(5)   American Oil Co. 1967. *Safe Operation of Refinery Steam Generators and Water Treating Facilities.* Safety Booklet No. 10. Chicago.
(6)   Noll, D. E. 1964. Factors that determine treatment for high-pressure boilers. *Proc. Amer. Power Conf.* 26:753.

Chapter 2.

Objectives in Treating Water for Boilers

Before proceeding with detailed discussions of the various processes for conditioning water for steam generators it may be helpful to summarize these processes and the reasons for carrying them out. The principal objectives in conditioning boiler water are to prevent scaling, corrosion, and contamination of steam by salts in the boiling water. Both physical and chemical methods are used to achieve these objectives. They are described briefly in the following sections, then considered in more detail in the next three chapters.

2.1 PREVENTION OF SCALING IN BOILERS

The term scale describes a continuous, adherent layer of foreign material formed on the water side of a surface through which heat is exchanged. By adding certain chemicals the growth of scales can be inhibited and the insoluble particles can be dispersed in the recirculating water and removed by blowdown. Should the particles come out of suspension, however, they can accumulate as sludges in quiet sections of a boiler. Deposit is a rather general term applied to more-or-less loose accumulations often found in less turbulent sections of boilers and water-treating systems. Scales are objectionable because of their insulating effect. In a boiler tube, for instance, they cause overheating and eventual failure of the metal. Deposits often cause plugging in critical areas such as waterwalls, waterwall headers, in blowdown lines, and in gauge glasses.

The coefficients of thermal conductivity of several metals and of some compounds of which scales may be composed are tabulated in Table 2.1.

TABLE 2.1

Thermal Conductivities of Various Materials

	$\lambda \times 10^3$ *
Metals	
Copper	920
Carbon steel	110
Bessemer steel	98
Scales	
Aluminum oxide, fused (Al_2O_3)	8.0
Analcite ($Na_2O \cdot Al_2O_3 \cdot 4SiO_2 \cdot 2H_2O$)	3.0
Calcium carbonate ($CaCO_3$)	2.2
Calcium phosphate [$Ca_3(PO_4)_2$]	8.6
Calcium sulfate ($CaSO_4$)	3.1
Ferric oxide (Fe_2O_3)	1.4
Magnesium oxide (MgO)	2.7
Magnesium phosphate [$Mg_3(PO_4)_2$]	5.1
Magnetite (Fe_3O_4)	6.9
Porous scales	0.2
Quartz glass (SiO_2)	3.6
Serpentine ($3MgO \cdot 2SiO_2 \cdot 2H_2O$)	2.4

* λ = g-cal/(s)(cm^2)(C/cm)
 Note: Btu/(h)(ft^2)(F/in.) $\times$ 3.44 $\times$ 10^{-4} = λ

Copper has the highest thermal conductivity of any common metal, so is an ideal material for equipment used for exchanging heat. Its tensile strength, however, is too low for use at high pressures. Boiler tubes are therefore fabricated of carbon steel, although its thermal conductivity is only about one-twelfth that of copper. Note, on the other hand, that the thermal conducitivity of carbon steel is nearly 80 times that of ferric oxide. Thus, when scales form on the inner surfaces of boiler tubes it is not possible to maintain the rate of heat transfer for which the boiler was designed.

It is sometimes said that a thin layer of calcium carbonate, called an eggshell coating, should be maintained within a boiler to protect the surfaces from corrosion. It is impossible to lay down a uniformly thin scale, however, because the thickness of the scale depends upon the amount of heat being transferred, which is not the same in all sections of a boiler.

Any scale in a boiler is undesirable. Scale-forming elements that should be removed from make up water for boilers include calcium, magnesium, iron, silicon, and aluminum.

a. Types of Scale

Many different mineral structures have been identified in boiler scales by the methods of x-ray diffraction, electron diffraction, and polarizing microscopy. Examples of silicate scales are: acmite, $Na_2O \cdot Fe_2O_3 \cdot 4SiO_2$; analcite, $Na_2O \cdot Al_2O_3 \cdot 4SiO_2 \cdot 2H_2O$; pectolite, $Na_2O \cdot 4CaO \cdot 6SiO_2 \cdot H_2O$; serpentine, $3MgO \cdot SiO_2 \cdot 2H_2O$; sodalite, $Na_2O \cdot 3Al_2O_3 \cdot 6SiO_2 \cdot 2NaCl$; and xonotlite, $5CaO \cdot 5SiO_2 \cdot H_2O$. When phosphate is used for internal treatment, ferric phosphate, $FePO_4$, basic magnesium phosphate, $Mg_3(PO_4)_2 \cdot Mg(OH)_2$, and hydroxyapatite, $Ca_{10}(PO_4)_6(OH)_2$, may also be encountered, as well as the more common anhydrite, $CaSO_4$, and aragonite, $CaCO_3$. As noted before, the presence of these and other scales impedes the circulation of water and reduces heat transfer, both of which cause overheating and failure of tubes.

b. Mechanism of Scale Formation

Scales and deposits form because the compounds of which they are composed are insoluble under the conditions prevailing in the boiler. Two factors combine to make calcium salts especially troublesome: certain anhydrous calcium salts, notably the sulfate, decrease in solubility as temperature and pressure increase, whereas increasing temperature shifts the equilibrium of the following reaction to the right, causing $CaCO_3$ to precipitate:

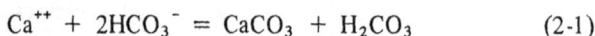

$$Ca^{++} + 2HCO_3^- = CaCO_3 + H_2CO_3 \qquad (2\text{-}1)$$

In addition, hydrolysis of excess bicarbonate increases the concentration of hydroxyl ion, precipitating $Mg(OH)_2$, the solubility product of which is 5.5×10^{-12}. The solubility of $CaSO_4$ decreases rapidly with increasing temperature, producing an extremely hard, adherent coating on boiler tubes, especially in locations where heat flux is high. The compositions of several scales containing aluminum, magnesium, calcium, and silicate are given above. Analcite and acmite, which form at high temperature, are invariably found beneath sludges of hydroxyapatite or serpentine, or under porous deposits of iron oxides. Occasionally other extremely insoluble iron or magnesium silicates are also encountered, and now and then α-quartz, SiO_2, appears, usually originating from colloidal silica, finely divided silt, or sand in the feed water.

Accumulations in boiler drums are most often in the form of mud or sludge. When oil is present as a contamination in boiler water, loose scales may form, particularly in water-wall tubes. Oil serves as a nucleus and binder for scaling at hot spots, although these scales are often merely baked mud that is easily dislodged by hammering the tubes. The "oil balls" found in steam drums and water-wall headers are typical formations in turbulent sections; they are especially common in steam drums, where they are formed by the rolling motion of water.

c. External Chemical Treatments

Obviously, the most effective method for preventing scaling is to eliminate scale-forming elements from the feed water, or to transform them by some means into an innocuous form. The methods for doing this are conveniently classified as external and internal treatments.

Chemical Softening. The treatments of water that are accomplished outside of the boiler are referred to as preboiler, or external treatments. The processes for removing calcium and magnesium ions from water are called softening, and are of signal importance in preventing scales. It is apparent from the equation

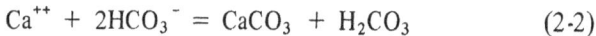

$$Ca^{++} + 2HCO_3^- = CaCO_3 + H_2CO_3 \qquad (2\text{-}2)$$

that calcium is precipitated if carbonic acid is neutralized by adding an alkaline reagent. If an excess of alkali is added, magnesium hydroxide also precipitates and the total hardness of the water is reduced. Lime is the alkaline reagent most often used because its cost is low and it is relatively easy to handle; the process is called lime softening. If noncarbonate calcium and magnesium hardness are present, i.e., if the M-alkalinity is less than the calcium hardness, sodium carbonate is also added to yield the following reactions:

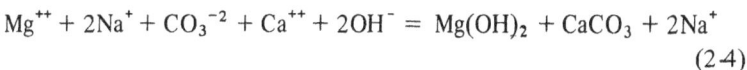

$$Ca^{++} + 2Na^+ + CO_3^{-2} = CaCO_3 + 2Na^+ \qquad (2\text{-}3)$$

$$Mg^{++} + 2Na^+ + CO_3^{-2} + Ca^{++} + 2OH^- = Mg(OH)_2 + CaCO_3 + 2Na^+ \qquad (2\text{-}4)$$

Filter alum $(Al_2(SO_4)_3 \cdot 18H_2O)$ is often used to improve the results of softening by the lime-soda process. Aluminum hydroxide formed by the hydrolysis of aluminum ion coagulates precipitates of calcium carbonate and magnesium hydroxide, yielding an effluent of lower turbidity and magnesium content than is otherwise obtained.

Softening by Cation Exchange. The removal of calcium and magnesium ions by cation exchange is commonly called zeolite softening, from the reaction characteristic of the mineral zeolites. The latter are hydrous sodium aluminum silicates in which the sodium is labile and exchangeable for calcium and magnesium ions flowing over or through the mineral. The exchange reaction is

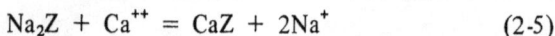

$$Na_2Z + Ca^{++} = CaZ + 2Na^+ \qquad (2\text{-}5)$$

Installations that require little treated water for make up, or systems that are operated at very high pressure, can make use of totally demineralized water. In this type of treatment the cation exchange resin is in the acid form, while the anion exchange resin, which removes negative ions, is in the hydroxide form. Examples of the reactions are

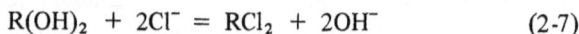

$$H_2Z + Ca^{++} = CaZ + 2H^+ \qquad (2\text{-}6)$$

$$R(OH)_2 + 2Cl^- = RCl_2 + 2OH^- \qquad (2\text{-}7)$$

Equivalent amounts of hydrogen and hydroxyl ions are produced that neutralize each other.

Silica Reduction. The concentration of silica in high-pressure systems can be conveniently controlled by modifying the methods of lime softening already mentioned. When using the cold lime softening process, ferric sulfate is added to produce a floc of hydrous ferric oxide that adsorbs silica. If a hot process is used for softening, magnesium hydroxide is more effective for reducing the concentration of silica. The most efficient removal of silica is achieved by anion exchange using a strong-base resin; the concentration of silica in the effluent is less than 10 ppb SiO_2. All of the methods summarized here are considered in detail in Chapter 4.

d. Internal Chemical Treatments

Precipitants. The amount of hardness that can be tolerated in feed water decreases as the pressure of the boiler increases, but in any case calcium and magnesium are potential formers of scale, so further treatment is required. Alkaline earth scales are prevented by adding phosphate to the boiler water; this precipitates both calcium and magnesium in a soft dispersed form. The precipitate formed by calcium and orthophosphate is usually represented as the normal phosphate, $Ca_3(PO_4)_2$. Attempts to precipitate this salt in the laboratory, however, invariably produce hydroxyapatite, the formula of which can be written in various ways including $Ca_{10}(PO_4)_6(OH)_2$, $3Ca_3(PO_4)_2 \cdot Ca(OH)_2$, and $Ca_5(PO_4)_3OH$. A consideration of the solubility products $(Ca^{++})(CO_3^{--}) = 4.8 \times 10^{-9}$, $(Ca^{++})^3(PO_4^{---})^2 = 1.3 \times 10^{-32}$, and $(Ca^{++})^5 (PO_4^{---})^3 (OH^-) = 3 \times 10^{-58}$ indicates that the basic salt forms in boiler water. Magnesium forms similar salts such as $Mg_3(PO_4)_2 \cdot Mg(OH)_2$ and $Mg_5(PO_4)_3(OH)$, which are undoubtedly much less soluble than $Mg(OH)_2$.

Magnesium salts are sometimes added to boilers operated at low pressure to precipitate magnesium silicate. This salt separates as a flocculent precipitate that can be removed by blowdown. Also, soda ash, Na_2CO_3, is used in low-pressure boilers fed with water containing 20–75 ppm of hardness to precipitate $CaCO_3$ and $Mg(OH)_2$.

Conditioners. Unless the amount of material precipitated in the boiler is small, it is necessary to add dispersants to prevent precipitated crystals from growing into large aggregates. These dispersants act by coating the particles to form a clarified colloidal solution that can be controlled by continuous blowdown; they are said to fluidize sludge. Older conditioners include tannins, lignins, sulfonated lignins, and starch. In modern practice polyacrylate and polymethacrylate are replacing the so-called natural organic polymers.

Chelants. The commonly used chelants are the tetrasodium salt of (ethylenedinitrilo)tetraacetic acid, $(NaOOCCH_2)_2NCH_2CH_2N(CH_2COONa)_2$, and the trisodium salt of nitrilotriacetic acid, $(NaOOCCH_2)_3N$. These are customarily referred to as EDTA and NTA, respectively. They form soluble complexes with various cations through the formation of stable ring structures. Calcium and magnesium form chelonates with both EDTA and NTA that are soluble and thus remain in solution. This makes for an exceptionally clean boiler, but the dosage of chelant must be carefully controlled for severe corrosion can occur if the excess of chelant is too large, particularly in the presence of oxygen. Chelants intended for treating boiler water are usually sold in formulations that also contain dispersants and antifoaming agents. The treatments mentioned in this section are described more fully in Chapter 5.

2.2 PREVENTION OF CORROSION IN BOILERS

Noll[1] has summarized the more common causes of corrosion failures peculiar to boilers and auxiliary steam generating equipment. Scattered pitting in the presence of oxygen is sometimes observed at the water line in the steam drum and in the downcomer tubes of boilers. Economizers, on account of their high temperature, are also particularly susceptible to corrosion by oxygen. Corrosion of iron and copper in condensate systems leads to the formation of porous deposits under which salts in boiler water concentrate and damage the underlying steel. Then too, even in the absence of deposits, caustic gouging can occur owing to the concentration of sodium hydroxide, particularly in places where the rate of heat transfer is unusually high. Other possibilities are the corrosion of stressed metal resulting from improper welding, and crevice corrosion under gaskets and backing rings. The severity of these effects can be mitigated to some extent by reducing the concentration of oxygen and free alkali, and by eliminating corrosion products introduced from the preboiler system.

a. The Mechanism of Corrosion

Slater and Parr[2] have shown the mechanism of pitting in a metallic surface produced by a bubble of air. Two stages in the formation of a pit are represented in Fig. 2.1. The two chemical half-reactions are

Cathodic

$$2e^- + \tfrac{1}{2}O_2 + H_2O = 2OH^- \tag{2-8}$$

Anodic

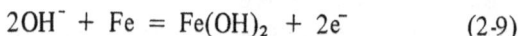

$$2OH^- + Fe = Fe(OH)_2 + 2e^- \tag{2-9}$$

Corrosion is the oxidation of metal by some oxidizing agent in the environment. The area over which the metal is oxidized is called the anode; that at which the oxidizing agent is reduced is called the cathode. These areas are necessarily separated, but usually are not far apart. As corrosion proceeds, electrons flow between these areas through the metal, while ions migrate through the solution. This system constitutes an electrochemical cell.

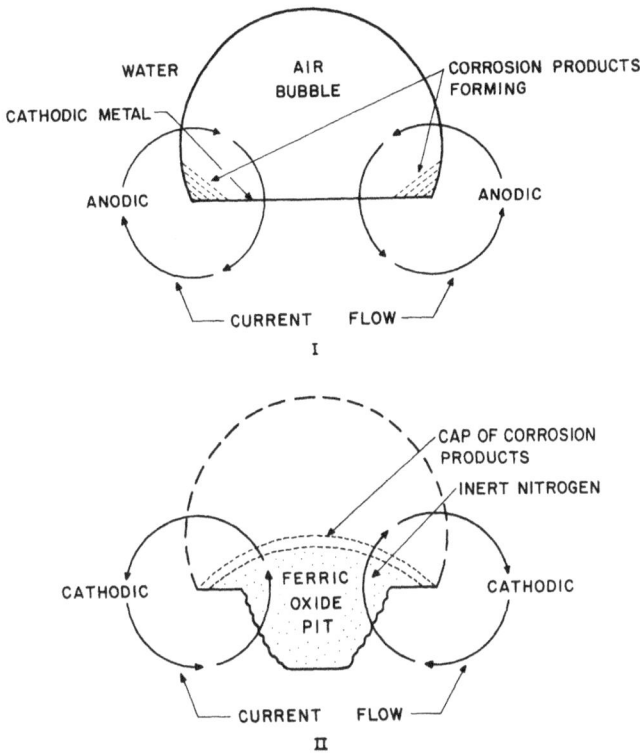

Fig. 2.1 Two stages of air bubble pitting

In boilers, the oxidation of iron is accomplished by the reduction of hydrogen ions supplied by the hot water.

$$3Fe + 4H_2O = Fe_3O_4 + 4H_2 \qquad (2\text{-}10)$$

In case of acidic water the reaction is

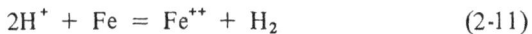

$$2H^+ + Fe = Fe^{++} + H_2 \qquad (2\text{-}11)$$

The reaction of Eq. (2-10) is self-limiting on account of the barrier of Fe_3O_4 that forms on the surface of the metal. The reaction in Eq. (2-11), on the contrary, continues until the supply of hydrogen ions is depleted— in boilers, only slightly impeded by polarization. Both reactions are opposed, however, by an irreversible potential called the hydrogen overvoltage, which is markedly affected by the condition of the surface of the metal. Agents that raise the overvoltage oppose, or inhibit, corrosion, whereas conditions that facilitate the escape of hydrogen gas from the cathodic surface (roughness, debris, etc.) augment corrosion. Other factors in localized corrosion include the presence of more than one metal, e.g., copper and iron; discontinous scale; and improper welding. Table 2.2 lists some typical anodic and cathodic couples.

TABLE 2.2

Anodic and Cathodic Couples

Anodic	Cathodic
Stressed metal	Unstressed metal
Low oxygen concentration	High oxygen concentration
Clean metal	Fouled metal
Iron	Copper

b. **Control of Corrosion**

Removal of Oxygen. Oxygen is introduced into boilers dissolved in the feed water. When this water enters the steam drum most of the oxygen flashes into the steam space, producing characteristic pitting at the water line and in the vicinity of the discharge of the feed line. In addition, Grabowski[3] has reported severe local corrosion, pinhole failures, and pitting in rear furnace-wall tubes in high-pressure boilers, attributable to attack by dissolved oxygen.

The concentration of dissolved oxygen in feed water should be less than 0.03 ppm, and preferably less than 0.005 ppm in water for high-pressure boilers. Cold water saturated with air contains about 10 ppm of oxygen. This can be reduced to 0.3–0.7 ppm in an open heater, and to about 0.01 in a spray-type deaerating heater. Finally, the concentration of oxygen is reduced to zero by chemical scavengers. Sodium sulfite is commonly used in boilers operating at less than 900 psi, while hydrazine is the reducing agent of choice at higher pressures.

Clarification of Boiler Water. The presence of porous deposits on the water side of boiler tubes leads to serious corrosion, especially when there is free alkali in the water. Various conditioners are added to disperse insoluble material and prevent the accumulation of sludge on surfaces where heat is transferred. Tannins, starch, sulfolignins, and quebracho have been used for many years as dispersants; more recently polyacrylates and other synthetic polymers have come into use.

Small amounts of particulate matter consisting of finely divided oxides of copper and iron often contaminate condensate. These oxides, in addition to causing foaming, deposit on the boiler tubes at a rate proportional to the heat flux; this rate of deposition increases rapidly above 850 psi. The presence of these deposits causes overheating of the tubes and sometimes ductile gouging. Direct reaction of steel with particles of ferric oxide is also possible.

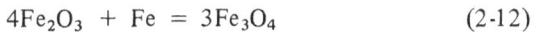

$$4Fe_2O_3 + Fe = 3Fe_3O_4 \qquad (2\text{-}12)$$

The Effect of pH. The metallic surface on the water side of a boiler tube is naturally protected by a thin film of magnetite formed as shown in Eq. (2-10) by the action of hot water on steel. Ideally, there is no further oxidation of metal after this protective layer is formed. The minimum rate of corrosion is realized at pH 11-12. At lower pH values hydrogen ions are discharged, whereas at values greater than 12 the magnetite layer thickens, peptizes to some extent, and is made porous by the diffusion of ions from the underlying metal. Above pH 14 the protective film is destroyed rapidly, and also above 500 psi hydrogen may diffuse into the metal, blistering and weakening it severely. Hydrogen atoms react with carbon in steel forming methane, CH_4; the pressure generated in this process of internal decarburization causes fissures along the grain boundaries.

Condenser leaks are a common cause of low pH values in marine boilers; a variety of mishaps can introduce acids into industrial boilers. The recirculation of a small amount of the alkaline boiler water through the feed pump has been recommended, but this practice can lead to plugging of feed lines, economizers, and feed water preheaters by insoluble phosphates.

Control of Alkalinity. Carbonates and bicarbonates decompose in hot boiler water to carbon dioxide and hydroxyl ion. The former escapes with the steam, while the latter remains in the water.

$$CO_3^{--} + H_2O = 2OH^- + CO_2 \qquad (2\text{-}13)$$

The hydroxyl ion formed by the reaction in Eq. (2-13) produces a pH of

11-12 in the boiler water; this method of alkalinity control is called the free caustic, or caustic reserve method. Under these conditions it is essential to measure and limit the total alkalinity of the boiler water to prevent excessive surface tension and concomitant carryover. The total alkalinity is kept at 10-20 percent of the total dissolved solids concentration, and can be regulated, for example, by increasing or decreasing the dosage used for chemical softening, or by adjusting the rate of addition of acid ahead of cation exchange softeners. High levels of hydroxide alkalinity foster thin, hard, dense, adherent layers of Fe_3O_4 on iron surfaces, but at low concentrations the layer of magnetite is porous. In the presence of porous deposits hydroxyl ions tend to concentrate between the metal and the deposit, leading to the typical corrosion pattern of caustic gouging. In highly concentrated alkali the protective film of magnetite dissolves, forming a mixture of ferrite and hypoferrite ions.

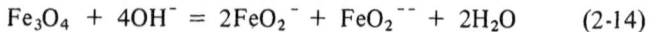

$$Fe_3O_4 + 4OH^- = 2FeO_2^- + FeO_2^{--} + 2H_2O \qquad (2\text{-}14)$$

This then exposes the underlying metal to concentrated alkali. When this localized concentration reaches about 40 percent the evolution of hydrogen begins, accompanied by the development of wide deep pits in the surface of the steel. When applying the caustic reserve system it is, therefore, important to keep the boiler clean and free from sludge. Attack is especially severe, of course, under high heat flux.

$$Fe + 2OH^- = FeO_2^{--} + H_2 \qquad (2\text{-}15)$$

Another method of controlling alkalinity is mentioned in Section 2.3, below, and is discussed in more detail in Chapter 5.

2.3 PREVENTION OF STRESS CORROSION CRACKING

Hydroxyl ion in contact with metal under stress causes deterioration and eventual failure of the metal by a process of embrittlement. This was common in riveted boilers, but in recent years, with the advent of welded, stress-relieved boilers, the phenomenon has become rare.

a. Mechanism of Caustic Embrittlement

The principle factors in caustic embrittlement are: localized strains, impurities in grain boundaries, stresses caused by improper welding, and seepage or leaking of boiler water through rivet holes or seams leading to localized concentration of free alkali. Stress corrosion cracking is predominantly intergranular and occurs only when steel is stressed beyond its elastic limit (or yield point) while in contact with a solution containing not less than the equivalent of 7.5 percent NaOH. Purcell and Whirl[4] have found that silicates equivalent to at least 0.16 percent SiO_2 are also necessary.

There has been considerable controversy over the mechanism of caustic embrittlement, but Tajc[5] has proposed a theory that takes into account the more important known factors. Namely, that intergranular corrosion is selective, that the concentration of hydroxyl ion is critical, and that silicates accelerate stress corrosion cracking. The theory assumes that cracking is controlled by surface phenomena. Hydroxyl ion peptizes the protective film of magnetite, and silicate acts as an inhibitor of general corrosion. Thus, the grains are protected at the expense of the grain boundaries, and cracking along the boundaries ensues. Hydrogen apparently does not figure in the mechanism.

b. Methods of Control

The most effective method for preventing corrosion of stressed metal is the coordinated phosphate-pH program proposed by Purcell and Whirl.[4] Briefly, it is a system for controlling alkalinity by coordinating pH and the concentration of the phosphate ion in accordance with a graph relating pH and the concentration of phosphate ion in solutions of pure trisodium phosphate. By hydrolysis, these solutions maintain sufficient alkalinity to inhibit corrosion, yet do not deposit free sodium hydroxide upon evaporation. The method is discussed further in Chapter 5.

Older methods include adding inhibitors such as sodium nitrate, quebracho extracts, and lignin sulfonates. Many years ago Parr and Straub[6] observed that certain waters that were high in bicarbonate and low in sulfate caused stress corrosion cracking. This fostered the belief that the ratio of sulfate to alkalinity is significant in controlling caustic embrittlement; subsequently this was found to be untrue.

2.4 PREVENTION OF STEAM CONTAMINATION

A most important objective in operating boilers is to produce clean, dry steam. Water droplets, even a small number, erode turbine blades severely, and can be destructive to other machinery. For instance, if water enters the drive cylinder of a reciprocating pump, the incompressible liquid may stop the piston short of the cylinder head, creating an enormous stress that is likely to crack the head. When steam is contaminated by boiler water, scales or deposits composed of the solids dissolved in the water form on superheater tubes, which reduces heat transfer, and on turbine blades, which impairs efficiency. Boiler water can be introduced into steam in several ways. Priming is a term that describes the violent surging of water into the steam outlet as a result of improper design, too high a firing rate, or fluctuations in steaming rate. Stabilized bubbles of steam that have little tendency to coalesce as they rise through the water produce foam; this can be caused by carrying too high a concentration of alkali or salts in the boiler water, or by various contaminants. Finally, a special type of contamination arises because of the tendency of silica to steam distill and form hard, adherent deposits on turbine blades.

a. Priming

Although priming is not influenced by chemical treatment, a short discussion of the phenomenon may be of interest. Fig. 1.1 represents the essentials of a thermal circulation loop in a water-tube boiler. If the device is filled with water, and heat is applied as shown, the water in the riser is warmed, its specific gravity decreases, and the weight of cold water in the downcomer displaces the warm water upward in the riser. Circulation thus commences with the water moving up the riser and down the downcomer. When the temperature of the water in both legs reaches 212 F, circulation ceases until the latent heat of vaporization (970 Btu/lb) has been absorbed, and steaming begins. Latent heat is the energy absorbed or released in the change of state from liquid to gaseous, or from gaseous to liquid.

Steam formed in the risers has some lifting effect, but circulation resumes primarily because the columns of water in the downcomers are heavier than those containing steam in the risers. As pressure increases, the size of the steam bubbles decreases and the densities of the solutions in the risers and downcomers approach each other—the density of steam increasing and that of water decreasing until at the critical pressure, 3203.6 psi, they are equal at 19.8 lb/ft^3. Downcomers, therefore, must be made longer as design pressure rises; in a 1400-psi stationary boiler the steam drum may be 120 ft above the firing floor.

At pressures exceeding 2000 psi, positive circulation, i.e., an external pump, is advisable to force water and steam through the boiler circuits; above 2650 psi positive circulation is mandatory. In this arrangement, much more lower head is required than for natural circulation. Also, much higher heat absorption rates can be achieved (200,000-600,000 Btu/h/ft^2) with positive circulation.

If the last pass of downcomers in a boiler develops steam bubbles because of high firing rate, circulation ceases momentarily and the boiler primes. That is, there is a violent upheaval of water within the boiler evidenced by a surge in the water glass. Liquid water is thrown into the steam outlet, and the risers are starved for water at a critical time. To assure proper cooling, water must move through the tubes at 2-5 ft/s with the circulation rate 5-20 times the evaporation rate. A sudden increase in steam load on a boiler momentarily decreases the pressure causing the water level to rise; priming sometimes occurs simultaneously. Other factors contributing to priming include too high or too low a water level, uneven firing distribution, wide swings in steam load, too high a rate of steaming, and occasionally, improper design.

b. Foaming

Foaming in boiler water is caused by high concentrations of dissolved or suspended solids, excessive alkalinity, and by saponifiable oils that form soaps. Foaming can be controlled to some extent by adding tannins and lignins, but better results are obtained with special inhibitors that are unsaponifiable and stable toward hydrolysis. The most efficient antifoams are polyamides and polyoxyalkylene glycols. As both of these are insoluble organic compounds that weaken the skin of a steam bubble, they must be well dispersed in the boiler water for maximum effect.

c. Volatilization of Silica

At pressures exceeding 500 psi, silica steam distills and forms extremely insoluble and adherent deposits on the inside of superheater tubes and on turbine blades. The volatility of silica is reduced by maintaining a high free alkalinity and by limiting the concentration of silica in the boiler water. If the concentration of silica in steam does not exceed 0.02 ppm, turbine deposits form slowly and are not troublesome. This matter is discussed in more detail in Section 5.5 and Section 6.2b.

REFERENCES

(1)   Noll, D. E. 1958. Limitations on chemical means of control-
      ling corrosion in boilers. *Corrosion* 14:541t.
(2)   Slater, I. G. and Parr, N. L. 1950. Marine boiler deterioration.
      J. Amer. Naval Engineers 62:405.
(3)   Grabowski, H. A. 1955. Corrosion of steel in boilers; attack
      by dissolved oxygen. *Trans. Amer. Soc. Mech. Engr.* 77:433.
(4)   Purcell, T. E. and Whirl, S. F. 1943. Protection against caus-
      tic embrittlement by co-ordinated phosphate-pH control.
      Trans. Electrochem. Soc. 83:343.
(5)   Tajc, J. A. 1937. A thermodynamic and colloidal interpreta-
      tion of published studies of corrosion cracking of stressed mild
      steel in water solutions. *Proc. Amer. Soc. Testing Materials*
      37, Part II:588.
(6)   Parr, S. and Straub, F. 1926. The cause and prevention of
      embrittlement of boiler plate. *Proc. Amer. Soc. Testing Ma-
      terials* 26, Part II:52.

Chapter 3.

Physical Methods for Improving Water Quality

Five physical methods for improving the quality of water for boilers are considered in this chapter: sedimentation, filtration, deaeration, deoiling, and blowdown. Sedimentation, more often than not, is the first step in the clarification of a water supply. Filtering follows several different processes in water treatment, including sedimentation, coagulation, and lime-softening. Deaeration removes oxygen and carbon dioxide, which if left in boiler feed water causes corrosion in feed water preheaters, boilers, and steam-condensate systems. Oil is removed from feed water to prevent foaming and film-boiling in boilers. Blowdown refers to the continuous or intermittent draining of a portion of the boiler water to avoid undue concentration of dissolved salts within the boiler and, thus, can be considered a method of conditioning the boiler water.

3.1 CLARIFICATION

Water from lakes, rivers, and streams usually contains suspended matter, as well as colloidal particles, including colored organic material. In general, water from these sources must be treated with chemicals that induce coagulation and flocculation, although in some instances simply allowing the solids to settle is sufficient to make water suitable for some industrial uses.

a. Sedimentation

The concentration of suspended mud, silt, and clay in natural waters can be reduced to about 10 ppm by settling in still ponds, or by slowly flowing the raw water through a system of settling basins. This method is used infrequently in industrial water treatment, as large areas are required for settling ponds. Also, the method is without effect on colloidal particles, which must first be coagulated, then flocculated into larger particles that settle under the influence of gravity. This process is discussed further in Section 4.1. As a rule, sedimentation is only useful as a preliminary step in clarifying exceptionally muddy water.

b. Filtration

With few exceptions, sedimentation must be followed by filtration to remove residual suspended solids. Industrial filters consist of layers of fine sand, or anthracite coal crushed and screened, resting on a graded gravel support; Nordell[1] gives typical specifications. Water passes downward through the graded beds that become increasingly coarse from top to bottom. A battery of filter units is required, the number depending upon the volume of water to be processed. If water is filtered by gravity, the usual rate is about 2 gpm/ft^2 of bed area. Pressure filters are more common in industry as they are not subject to air-binding, and a rate of 3 gpm/ft^2 of bed area can be achieved.

Filtration continues until loss-of-head gauges indicate plugging of the filter. When the pressure drop across the filter reaches 5-8 psi, the unit is taken off the header and cleaned by passing filtered or municipal water upward through the bed at 10-15 gpm/ft^2. Rotary surface washers are operated during the backwash to dislodge accumulated sludge and other fouling from the surface of the filter. At the same time the upward flow of water hydraulically grades the filter bed. Backwash water is customarily collected in a reclamation tank where the sludge is allowed to settle, after which the supernatant water can be reused. Finally, the filter unit is returned to service, after filtering to waste for a short time while the bed subsides.

3.2 DEAERATION

The greater part of the corrosive gases, carbon dioxide and oxygen, that are dissolved in water can be removed by deaeration. Vacuum deaerators are not efficient enough to be used for preparing feed water for boilers, but they are satisfactory for water to be used for condensing units in air conditioners, for the manufacture of ice and carbonated beverages, and for several other applications in which nearly complete removal of the gases is unnecessary, or perhaps undesirable. Pressure is reduced in these vessels by multistage 100-psi steam ejectors. Under these conditions the water boils cold, so there is negligible decomposition of bicarbonate.

Open heaters are suitable for deaerating feed for low-pressure steam generators, but for pressures above 400 psi spray-type deaerating heaters are commonly used. In these units exhaust steam at 1-10 psi, or steam from a low-pressure flash tank, heats the feed water in a primary heater and also scrubs the heated water. Hot water sprays downward through nozzles against a rising flow of steam that sweeps the liberated gases out through a vent at the top of the vessel, while the deaerated water collects in a storage section at the bottom. The temperature at the top of the scrubbing section should be at $(212 + 3 \times psi)$ F.

The vent is equipped with a condenser through which cold feed water flows to prevent excessive wastage of steam. Nordell[1] points out that 1000 gal of water containing 10 ppm of carbon dioxide and saturated with air at 50 F, releases about 5 ft^3 of gases in a deaerator at 212 F. Even though this large volume of noncondensable gas is swept out through the vent by steam, only a small plume of steam two to three feet high is wasted if the unit is provided with a vent condenser.

If exhaust steam from reciprocating machinery is used for deaeration, an oil separator consisting of a ribbed baffle enclosed in a welded steel shell should be installed in the steam supply line. Steam strikes the baffles, changes direction horizontally, and passes around the baffles and out through ports on either side. Droplets of oil adhere to the baffle, ultimately dripping from the bottom of it into a reservoir.

Table 3.1 shows the approximate concentrations of carbon dioxide and oxygen in the effluents from the three types of deaerators described. Specifications for oxygen concentration are often expressed as less than 0.007 ppm (0.005 ml/l) of oxygen, as this is the limit of detection by the Winkler method.[2] A newer method using indigo carmine,[3] however, can detect less than 5 ppb of oxygen.

Because of the volatilization of carbon dioxide and the thermal decomposition of bicarbonate, the pH of deaerated water is normally 8.5-9.5.

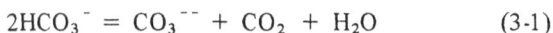

$$2HCO_3^- = CO_3^{--} + CO_2 + H_2O \qquad (3\text{-}1)$$

$$H^+ + HCO_3^- = CO_2 + H_2O \qquad (3\text{-}2)$$

The pH rises with the pressure and temperature of the deaerating steam.

TABLE 3.1

Efficiency of Deaerators

Type	Concentration in effluent	
	ppm CO_2	ppm O_2
Vacuum deaerator (cold)	5-10	0.4-1.4
Open heater	0	0.3-0.7
Spray-type deaerating heater	0	0.005-0.01

It is advantageous to maintain as low a concentration of bicarbonate as possible in feed water for boilers to reduce the concentration of carbon dioxide in the steam-condensate system. Much detailed information on the mechanical aspects of deaerators is included in a publication of the Permutit Company.[4]

3.3   OIL REMOVAL

Oil contamination may be dispersed, entrained, emulsified, or dissolved in water. Dispersed or entrained oil is distributed in macro-drops, either uniformly or nonuniformly, in the water. Oils with negligible solubility, which include paraffins, mineral oil, and uncompounded lubricating oils, usually appear as visible droplets. Some cylinder oils are compounded with saponifiable animal fats, most commonly raw degras, lard oil, or tallow, that can act as emulsifying agents that form very stable emulsions with steam condensate. Benzene, xylenes, and other mononuclear aromatics are appreciably soluble in water, but they volatilize rapidly with steam and do no harm in the boiler. They do, however, strip filming amines from the condensate system.

In boiler water, concentrations of oil up to 10 ppm can be coagulated and rendered innocuous by starch (50-100 ppm), or by polyacrylates (10-15 ppm). The effect of higher concentrations, however, is to make the surface of the tubes oily and nonwetting, which in turn leads to film boiling and overheating of the metal. If a contaminating oil is saponifiable, foaming is also likely. Oil serves as a nucleus and binder for scaling at hot spots, and it frequently agglomerates boiler sludge and suspended solids into "oil balls," especially in turbulent sections such as steam drums and water-wall headers, on account of the boiling of the water there. Also, loose scales containing oil are now and then discovered in water-wall tubes. It is most important that oil be completely removed from water before passing it through demineralizers or cation exchangers, for if oil is present it coats the spherical particles of resin, mechanically blocking exchange of ions.

The most satisfactory method for removing oil from condensate is to pass the water through a filter precoated with a hydrous oxide gel such as aluminum or ferric hydroxide, both of which have an affinity for oil. Oil is adsorbed on the surface of the gel, which is periodically flushed off of the supporting filter, then replaced with a fresh layer. A coating of pre-formed hydrous aluminum oxide gel supported on a bed of graded anthra-cite coal is a proven medium for removing moderate concentrations (10-50 ppm) of oil from condensates. The filter units are cleaned by backwash-ing with filtered hot condensate, or with fresh water heated to reduce the viscosity of the oil, while at the same time agitating the bed with an inter-nal rake, or a rotary surface washer. Backwashing, at 10-12 gpm/ft^2 of surface area, is initiated when the loss of head pressure reaches 4-5 psi, and is continued for 10 min. Filters in service are operated at 5-8 gpm/ft^2 of surface area.

Aluminum hydroxide gel is conveniently prepared by mixing filter alum, $Al_2(SO_4)_3 \cdot 18H_2O$, with soda ash, Na_2CO_3, in a separate mixer. Alumi-num ion is completely hydrolyzed in carbonate solutions to $Al(OH)_3$, but as aluminum hydroxide it is capable of acting either as a weak base or a weak acid, it is essential that the proportions of the two chemicals be accurately measured.

$$Al^{+++} + 3CO_3^{--} + 3H_2O = Al(OH)_3 + 3HCO_3^{-} \qquad (3\text{-}3)$$

If either an excess or a deficiency of sodium carbonate is used, soluble aluminum enters the boiler where it is prone to form exceedingly hard scales with silicate. Table 3.2 shows the pH, concentration of aluminum ion in solution, and the stability of the emulsion formed upon mixing vari-ous weights of sodium carbonate with 12 lb of aluminum sulfate in 48 gal of water.

These results show that aluminum is completely insoluble when sodium carbonate and alum are combined in such proportions that the pH of the gel after mixing falls between 5.9 and 7.1; this happens if 0.4-0.6 lb of

TABLE 3.2

Aluminum Hydroxide Gel for Precoat Filters

Chemical (lb/48 gal)		pH after mixing	Aluminum in solution (ppm)	% dispersed solids		
Filter alum	Soda ash			1 h	24 h	48 h
12	0	3.1	–	–	–	–
12	3	4.2	95	66	48	41
12	4	5.1	.07	70	68	66
12	5	5.9	nil	92	71	71
12	6	6.5	nil	93	75	75
12	7	7.1	nil	96	84	79
12	8	8.2	.01	99	88	85
12	9	9.0	.24	95	82	73

soda ash is used per pound of filter alum. Fig. 3.1 shows the daily requirement of alum for removing various concentrations of oil from oily water.
The figure is constructed on the assumption that 0.3 lb of alum is used per
pound of oil to be removed.

Air flotation separators are appropriate for removing large amounts of
oil (1200 ppm) from reclaimed water, while at the same time eliminating
the bulk of any suspended solids that might be present. In an induced air
flotation separator water flows into a pressurized cell where air is dispersed
in minute bubbles that are then collapsed by increasing the pressure. The
saturated solution passes into a coagulation chamber, the pressure is reduced, and as the air comes out of solution, bubbles form preferentially on

TABLE 3.3

Efficiency of Different Methods for Deoiling Water

Method	Oil in (ppm)	Oil out (ppm)
Precoat filters	50	< 1
Air flotation	60-1200	10-50
Air flotation with polymer	60-1200	5-10
Coalescing filters	10-50	3-7

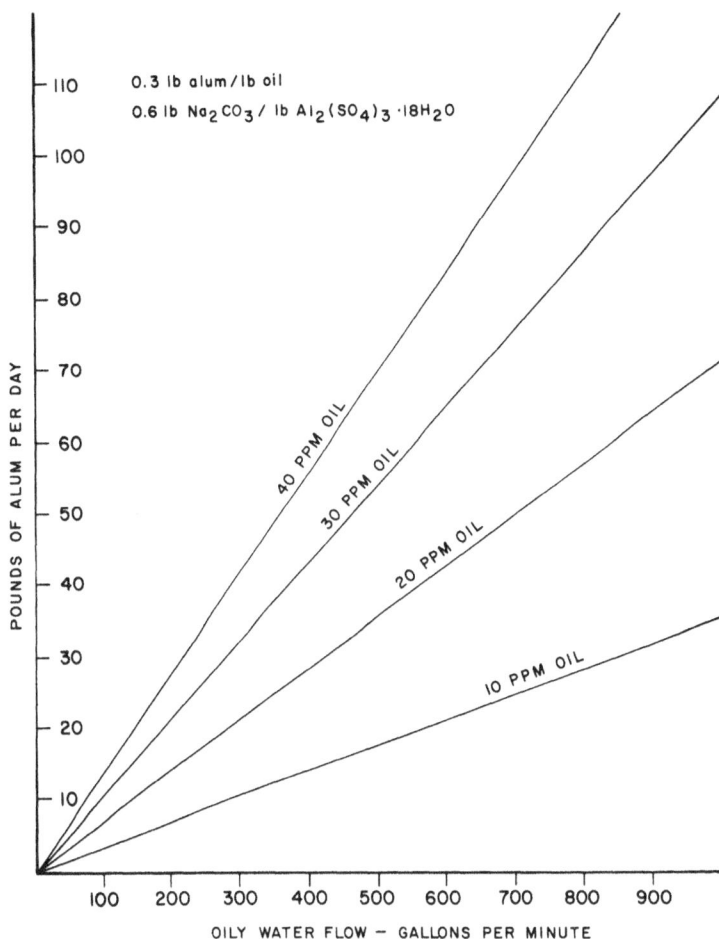

Fig. 3.1   Oily Water Flow (gal/min.)

nucleation sites provided by droplets of oil and particles of silt, lifting them to the surface of the vessel where they are skimmed. Nonionic polymers added to the water improve the efficiency of oil removal by this method, which is particularly suited for separating emulsified oil from water.

Forcing water through coalescing filters with very small pores ($1\text{-}40\ \mu$) reduces the concentration of oil contamination to something less than 10 ppm, but silt and other suspended solids must be absent, as they clog the filter cartridges rapidly. This process is best for oily water in which oil may be present in various forms, i.e., free, dispersed, or emulsified. Table 3.3 contains comparative figures showing the efficiencies of the available methods for removing oil from water for feeding boilers.

3.4 BLOWDOWN

As steam leaves a boiler, solids introduced in the feed water are concentrated in the water left behind in the boiler. If this concentration were allowed to continue, the less soluble components in the water would eventually crystallize on the internal surfaces and, in addition, the steam would become contaminated. In ideal operation the concentration of solids is allowed to reach the limits given in Table 3.4, after which the concentrated water is bled off at such a rate that the amount of solids entering in the feed water is exactly balanced by that removed in the bleed stream. This process is called continuous blowdown.

TABLE 3.4

Permissible Concentrations of Solids in Boilers

Pressure (psi)	Dissolved solids (ppm)	Suspended solids (ppm)	Total alka- linity (ppm)	Silica (ppm)
100	5000	500	900	250
200	4000	350	800	200
300	3500	300	700	175
500	3000	60	600	40
600	2500	50	500	35
750	2000	40	300	30
900	1000	20	200	20
1000	500	10	50	10
1500	150	3	0	3
2000	50	1	0	1

Suspended solids in the presence of iron, or when treated with coagulants, tend to collect as sludge in the lower mud drum of a boiler. This concentrated sludge can be blown out by opening briefly a hand-operated valve customarily provided for intermittent blowdown. Turbulence caused by opening the valve disperses sludge in the lower drum, so there is no point in leaving the valve open longer than 15 s. Herman and Gelosa[5] recommend the installation of a V-shaped angle iron over the opening of the bottom blowdown extending the entire length of the mud drum to remove sludge more efficiently. It is a dictum in boiler practice that water-wall headers should never be blown down, because the circulation of water through them is usually critical. In some of the older boiler designs, however, this can be done safely, and when this is the case, the accumulation of phosphate muds in the headers can be avoided by making several short "puff blows" once every eight hours. The boiler manufacturer should, of course, be consulted before initiating this inherently hazardous procedure.

The outlet for continuous blowdown should be above the level where the riser tubes enter the steam drum, because this is where the dissolved solids in the recirculating water are most concentrated. Chemicals for internal treatment (phosphate, sulfite, etc.) should be introduced above the downcomer tubes to prevent sludging on the hot risers, and also to promote mixing and reaction with salines in the entering feed water. If chemicals are injected near the blowdown outlet, short-circuiting and erroneous chemical analyses will be obtained. Concentrations of chemicals will be high in the blowdown water, but actually low in the recirculating water.

Continuous blowdown is the most effective method for controlling the amount of dissolved solids in the boiler water after the rate of withdrawal has been adjusted properly. If the rate of blowdown is too high, heat (fuel) and water are wasted; if too low, the permissible limits quoted in Table 3.4 will be exceeded. Percentage blowdown can be expressed in two ways:

In terms of water evaporated,

$$\text{Percent blowdown} = \frac{\text{total solids in feed} \times 100}{(\text{total solids in boiler} - \text{total solids in feed})}$$

In terms of feed water,

$$\text{Percent blowdown} = \frac{\text{total solids in feed} \times 100}{\text{total solids in boiler}}$$

The cycles of concentration, or blowdown ratio, is equal to the concentration of solids in the boiler water divided by the concentration of solids in the feed water. This is the number of times the feed water has been concentrated. It was pointed out in Section 1.4 that the blowdown ratio is equal to the feed rate divided by the blowdown rate.

$$f/b = R \qquad (3\text{-}4)$$

Therefore,

$$b \times 100/f = 100/R$$

and

$$\text{percent } b = 100/R \qquad (3\text{-}5)$$

Eq. (3-9) is a function of the form $y = a/x$, the graph of which is an equilateral hyperbola, discontinuous at $x = 0$. It is sometimes convenient to take the logarithm of both sides of Eq. (3-9), yielding

$$\log R + \log (\text{percent } b) = 2 \qquad (3\text{-}6)$$

which plots as a straight line on 2-cycle logarithmic graph paper. Such a plot is useful for reading percent b directly after R has been determined by chemical analysis.

REFERENCES

(1)   Nordell, E. 1951. *Water treatment for industrial and other uses.* New York: Reinhold Publishing.

(2)   Winkler, L. W. 1888. The determination of dissolved oxygen in water. *Ber.* 21:2843.

(3)   Amer. Soc. Testing Materials. 1977. *Standard Test Methods for Dissolved Oxygen in Water.* 1977 Annual Book of ASTM Standards, Part 31, Method D888-66 438.

(4)   *Water conditioning handbook.* 1954. New York: The Permutit Company.

(5)   Herman, K. W. and Gelosa, L. R. Apr., 1973. Water treating for heating and process steam boilers. *Power Eng.* 77 (4):54.

Chapter 4.

External Chemical Treatments

Treatments of water that are done outside of the boiler are called pre-boiler, or external treatments, the most important of which are coagulation by chemicals, softening, and the removal of silica. When preparing water for boilers operated at less than 150 psi, all necessary chemical treatment can be accomplished in a clarifier, but as pressure increases the quality of the feed water must also improve (see Table 6.11). The following sections describe in some detail the methods available for reducing suspended solids, softening water, removing silica, and the ultimate in purification of water by chemical methods, demineralization.

4.1 COAGULATION BY CHEMICALS

Aluminum hydroxide, formed by the hydrolysis of aluminum ion, coagulates precipitates of calcium carbonate, magnesium hydroxide, silt and other suspended solids, as well as colloidal organic materials, producing an effluent of lower turbidity than is otherwise obtained. In addition, total dissolved solids and total hardness are also reduced.

The coagulating action of filter alum ($Al_2(SO_4)_3 \cdot 18H_2O$) depends upon two factors. The first is that colloidal particles that cause color and turbidity are, in general, negatively charged. The addition of a positive ion to such a solution causes coagulation of the colloidal particles, the effectiveness increasing with the charge of the ion. Thus, the trivalent ions of aluminum and iron are most often used for clarification. The second factor is the hydrolysis of the aluminum ion in solutions of alkali, carbonate, or bicarbonate to form a gelatinous precipitate of hydrous aluminum oxide. Adsorption cf particles on the surface of the gelatinous precipitate of aluminum hydroxide also increases its efficiency as a clarifying agent. Black[1] has discussed the mechanism of coagulation by aluminum salts, as well as

38

the use of starch, gelatin, and synthetic polymers as aids to coagulation.

Sodium aluminate ($NaAlO_2$) is sometimes preferred in situations where it is undesirable either to decrease the alkalinity of the water being treated or to increase its content of sulfate:

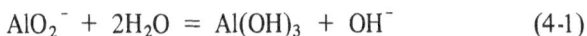

$$AlO_2^- + 2H_2O = Al(OH)_3 + OH^- \qquad (4\text{-}1)$$

Aluminate ion also reacts with carbon dioxide:

$$2AlO_2^- + CO_2 + 3H_2O = 2Al(OH)_3 + CO_3^{--} \qquad (4\text{-}2)$$

Commercial sodium aluminate is formulated with excess sodium carbonate and sodium hydroxide to prevent the precipitation of hydrous aluminum oxide in storage tanks and mixers.

When using alum as a coagulant a minimum dose of 5 ppm is needed to achieve the necessary supersaturation; the minimum methyl orange alkalinity after coagulation is 10 ppm $CaCO_3$. If the natural alkalinity is too low, 0.4 ppm of $Ca(OH)_2$, or 0.9 ppm of Na_2CO_3 should be added per ppm of alum used. Sodium aluminate yields about three times more aluminum than does alum, and in the long run is more economical to use if the water to be treated is not too alkaline.

When aluminum or ferric salts are used as primary coagulants the turbidity in the effluent from the clarifier can often be reduced by supplementing the inorganic coagulants with 0.1–0.5 ppm of a nonionic polymer. Cationic polymers with high charge density are primary coagulants that also neutralize the negative charges on colloidal particles and induce coagulation. As 0.3 to 3 ppm of a cationic polymer is generally sufficient to clarify water, the organic treatment may be more economical than inorganic coagulation. Anionic polymers are suited to processes that produce solutions containing positively charged particles such as lime-soda softening, which is discussed next.

4.2 SOFTENING

The processes for removing calcium and magnesium ions from water are called softening and are of signal importance in the prevention of boiler scale. Chemical softening using lime, or lime and soda ash, reduces total dissolved solids, and is applicable to water containing turbidity, iron, and organic matter. It does not reduce hardness to zero, however, so it cannot be used by itself to prepare make up water for boilers operated in excess of 200 psi. Also, a large volume of sludge is formed containing about 5 percent of solids, the disposal of which is expensive. Recirculation of a small amount ($\approx 1 \%$) of the sludge invariably improves floc formation. Cation exchange is the softening method of choice for clarified waters intended for boilers operated in the range of 300–1000 psi.

a. Cold Lime-Soda Process

It is apparent from the equation

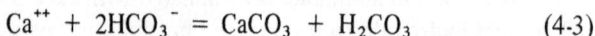

$$Ca^{++} + 2HCO_3^{-} = CaCO_3 + H_2CO_3 \tag{4-3}$$

that calcium is precipitated as the carbonate if carbonic acid is neutralized by adding an alkaline reagent. If an excess of alkali is added, magnesium hydroxide also precipitates and the total hardness of the water is reduced. Lime is the alkaline reagent most often used because of its low cost and relative ease of handling; the process is called lime softening. If the bicarbonate alkalinity (M-alkalinity) is equal to or in excess of the total hardness, the following reactions occur upon adding hydrated lime to the water:

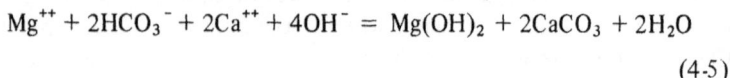

$$Ca^{++} + 2HCO_3^{-} + Ca^{++} + 2OH^{-} = 2CaCO_3 + 2H_2O \tag{4-4}$$

$$Mg^{++} + 2HCO_3^{-} + 2Ca^{++} + 4OH^{-} = Mg(OH)_2 + 2CaCO_3 + 2H_2O \tag{4-5}$$

In the reaction in Eq. (4-4) one equivalent of lime precipitates one equivalent of calcium hardness, while in the reaction in Eq. (4-5) two equivalents of lime remove one equivalent of magnesium hardness.

If noncarbonate calcium and magnesium hardness are present, i.e., if the M-alkalinity is less than the calcium hardness, sodium carbonate can be used to produce the following reactions:

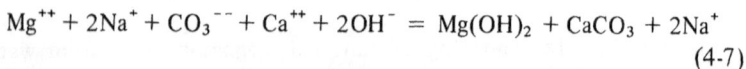

$$Ca^{++} + 2Na^{+} + CO_3^{--} = CaCO_3 + 2Na^{+} \tag{4-6}$$

$$Mg^{++} + 2Na^{+} + CO_3^{--} + Ca^{++} + 2OH^{-} = Mg(OH)_2 + CaCO_3 + 2Na^{+} \tag{4-7}$$

Eq. (4-6) indicates that one equivalent of noncarbonate calcium hardness requires one equivalent of sodium carbonate (soda ash), while in Eq. (4-7) one equivalent of noncarbonate magnesium hardness requires one equivalent of soda ash *plus* one equivalent of lime. This latter process is called lime-soda softening. The foregoing relationships are summarized in Table 4.1.

TABLE 4.1

Chemical Requirements for Lime-Soda Softening

	Equivalents required	
Hardness Type	$Ca(OH)_2$	Na_2CO_3
Carbonate		
Calcium	1	–
Magnesium	2	–
Noncarbonate		
Calcium	–	1
Magnesium	1	1

The characterization of hardness as carbonate or noncarbonate depends upon three variables: calcium hardness, magnesium hardness, and total alkalinity. Thus, it would appear that six different relationships might exist with respect to the relative values of these entities.

For example, if $CaH < M < MgH$

Carbonate calcium hardness	$= CaH$
Carbonate magnesium hardness	$= 2(M - CaH)$
Noncarbonate calcium hardness	$= 0$
Noncarbonate magnesium hardness	$= MgH - (M - CaH) = (TH - M)$

If the other five permutations are examined, however, it will be found that they reduce to two cases. Namely, $M > CaH$ and $M < CaH$.

Case A: $M > CaH$.

The distribution of hardness is that shown in the example above.

Case B: $M < CaH$.

Carbonate calcium hardness	$= M$
Carbonate magnesium hardness	$= 0$
Noncarbonate calcium hardness	$= (CaH - M)$
Noncarbonate magnesium hardness	$= MgH$

From Eqs. (4-4), (4-5), (4-6), and (4-7), and the equivalent weights $A = CaCO_3$, $B = Ca(OH)_2$, and $C = Na_2CO_3$, the dosages required for softening can be calculated for the two cases.

Case A: $M > CaH$.

Lime required:

$$CaH \times (B/A) + (M - CaH) \times 2 \times (B/A) + [MgH - (M - CaH)] \times (B/A)$$

or

$$[CaH + 2 \times (M - CaH) + MgH - (M - CaH)] \times (B/A) =$$
$$(M + MgH) \times (B/A)$$

$$(M + MgH) \times 37.1/50 = ppm\ Ca(OH)_2$$

$$0.742(M + MgH) = ppm\ Ca(OH)_2$$

$$0.560(M + MgH) = ppm\ CaO$$

Soda ash required:

$$[MgH - (M - CaH)] \times (C/A) = (CaH + MgH - M) \times (C/A)$$

or

$$(TH - M) \times (C/A) = ppm\ Na_2CO_3$$

$$1.06(TH - M) = ppm\ Na_2CO_3$$

Case B: $M < CaH$.

Lime required:

$$M \times (B/A) = M \times 37.1/50 = ppm\ Ca(OH)_2$$

$$0.742M = ppm\ Ca(OH)_2$$

$$0.560M = ppm\ CaO$$

Soda ash required:

$$(CaH - M) \times (C/A) + MgH \times (C/A) = ppm\ Na_2CO_3$$

$$1.06(TH - M) = ppm\ Na_2CO_3$$

If $CaH = M$, cases A and B are equivalent because $(M - CaH)$ and

$(CaH - M)$ are both zero and $(TH - M)$ is equal to MgH.

As an example of the dosage calculation, suppose one wants to soften water having the following analysis:

Total hardness, TH   = 103
Calcium hardness, CaH   = 68
Magnesium hardness, MgH   = 35
M-alkalinity, M   = 78

It is noted that TH $>$ M and that the noncarbonate hardness is $(TH - M)$. Then, as M $>$ CaH, the calcium carbonate hardness is equal to CaH. The magnesium carbonate hardness is $(M - CaH)$, and the magnesium noncarbonate hardness is $[MgH - (M - CaH)]$. As there is no calcium noncarbonate hardness (because CaH $<$ M), Eq. (4-6) does not apply. Therefore, Case A (M $>$ CaH) is to be used.

Lime required:

$$0.742(M + MgH) = \text{ppm Ca(OH)}_2$$

$$= 0.742(78 + 35)$$

$$= 83.8 \text{ ppm Ca(OH)}_2$$

Soda ash required:

$$1.06(TH - M) = \text{ppm Na}_2CO_3$$

$$= 1.06(103 - 78)$$

$$= 26.5 \text{ ppm Na}_2CO_3$$

b. Hot Lime-Soda Process

When the lime-soda process is carried out at normal temperatures calcium hardness can be reduced to about 35 ppm $CaCO_3$, with about 90 percent of the magnesium hardness remaining in solution. In the example just cited a total hardness of some 66 ppm $CaCO_3$ would be expected, but Wood[2] points out that in practice 85 ppm $CaCO_3$ is all that is achieved in cold process softening. If the process is carried out hot, the total hardness is reduced to about 20 ppm $CaCO_3$ on account of improved coagulation of the precipitated salts in the hot solution. By adding a small amount of phosphate (5-8 ppm residual after filtering) the total hardness is further reduced to about 2 ppm $CaCO_3$. Recirculating a small amount of sludge and adding a nonionic polymer improves the settling of the sludge and reduces the turbidity of the effluent water. Best results are obtained when the retention time is at least one hour; chemicals are fed as a slurry proportioned to the flow of water through the unit. The M-alkalinity from both the hot and cold processes is adjustable, but is normally 40-60 ppm $CaCO_3$.

The same calculation of dosage developed for cold lime-soda softening applies to hot softening, but the method of controlling the process is different. Any change in the composition of the raw water that effects a reduction in lime demand reveals itself by an increase in the value of $(2P - M)$.

In the softened water, if

$(2P - M) < 5$, increase lime
$(2P - M) > 15$, decrease lime
$[(2(M - P) - TH] < 20$, increase soda ash
$[(2(M - P) - TH] > 35$, decrease soda ash

$2(M - P)$ is the calcium carbonate equivalent of carbonate ion in the effluent.

These limits are embodied in Fig. 4.1, which is a dual graph in which the ordinate is either P-alkalinity or total hardness, and the abscissa is either M-alkalinity or $2(M - P)$. Furthermore, the graph is divided into nine regions corresponding to excess or deficiency of lime or soda ash and correct treatment. Care is required in using this chart to avoid confusing the two sets of coordinates. P-alkalinity and M-alkalinity govern lime dosage; $2(M - P)$ and total hardness govern soda ash. To illustrate the use of the graph, suppose the softened water from a hot process softener shows the following analysis:

P-alkalinity $= 40$
M-alkalinity $= 100$
Total hardness $= 80$
$2(M - P)$ $= 120$

The P- and M-alkalinity coordinates lie in region E, where an increase in lime is indicated. The Total hardness and $2(M - P)$ coordinates lie in region C, where a decrease in soda ash is called for. Therefore, the dosage of lime should be increased and that of soda ash decreased to approach the correct treatment region.

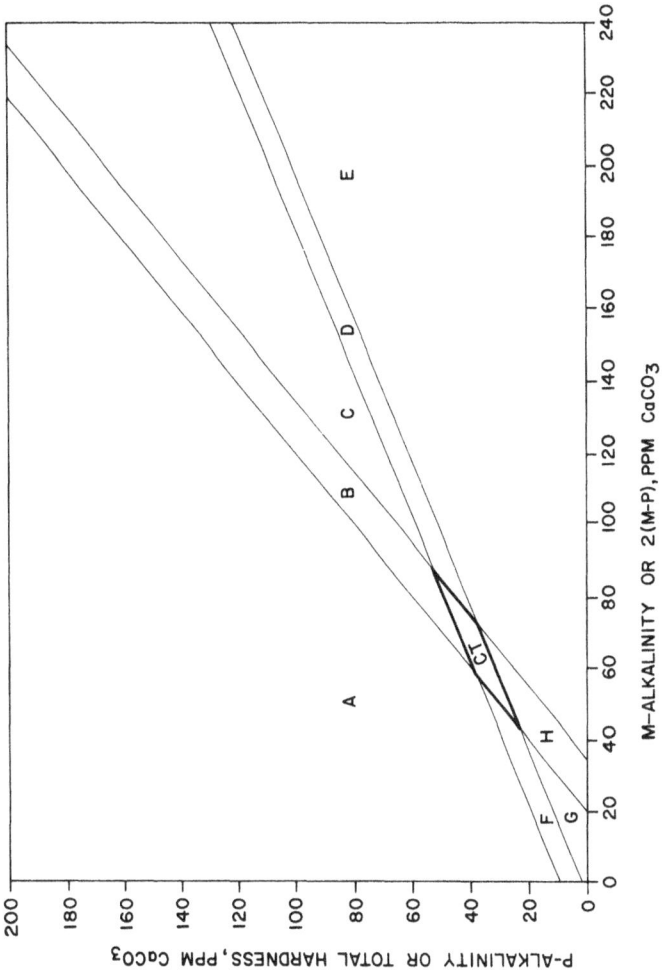

Fig. 4.1 Correct treatment chart for hot process water softening

c. Cation Exchange

The reaction characteristic of the natural zeolites (sodium aluminum silicates), in which sodium in the mineral is exchanged for calcium and magnesium in water, is called cation exchange, and is especially suited for removing so-called permanent hardness from water used for feeding boilers. Synthetic resins[3,4] have largely replaced the natural greensands and sulfonated coals formerly used for softening water. In Table 4.2 the capacities of several different types of exchangers are given with respect to the exchange reaction

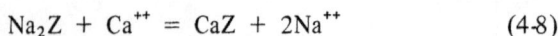

$$Na_2Z + Ca^{++} = CaZ + 2Na^{++} \qquad (4\text{-}8)$$

TABLE 4.2

Capacities of Various Cation Exchangers

Type	Average capacity meq/ml*	gr $CaCO_3/ft^3$
Natural zeolite (greensand)	0.16	3,500
Sulfonated coal	0.39	7,500
Synthetic zeolite	0.53	11,500
Phenolic resin	0.64	14,000
Sulfonated polystyrene	1.50	32,000

* 1000 meq/l = 21,850 gr $CaCO_3/ft^3$

When all of the active sites for exchange have been occupied, hardness begins to appear in the effluent. If the water being treated contains both calcium and magnesium ions, magnesium appears first in the effluent because, of the elemental ions having the same oxidation number (valence), the one with the higher atomic weight is most firmly held by the resin.

When the resin has become saturated, as indicated by the appearance of hardness in the effluent, it is regenerated by passing a solution of sodium chloride through the bed, after which the unit is rinsed with water. Advantages of using sodium chloride for regenerating include low cost, low molecular weight, high solubility of chlorides of calcium and magnesium, and ease and safety of handling.

A consideration of molecular weights shows that approximately 0.17 lb of sodium chloride is required to displace 1000 gr of equivalent $CaCO_3$ from an exhausted resin:

$$CaR_2 + 2Na^+ = 2NaR + Ca^{++} \qquad (4.9)$$

Two molecules of sodium chloride displace one molecule of calcium carbonate, or 117 lb of NaCl displaces 100 lb of $CaCO_3$. Thus, the stoichiometric requirement is 1.17 lb NaCl/lb $CaCO_3$. It is customary to express the capacity of ion exchangers in kilograms of calcium carbonate per cubic foot of resin; as 1 lb is 7000 gr, the theoretical requirement is 0.168 lb NaCl per kg $CaCO_3$. As would be predicted from the law of mass action, however, a large excess of sodium ion is required to force the regenerating reaction to the right.

In Table 4.3 the salt efficiency is tabulated at several different dosages of sodium chloride. It is seen that as the amount of salt increases, the capacity of the cation exchanger increases. There is a point, however, beyond which it becomes uneconomical to increase the amount of salt. The optimum efficiency is about 0.5 lb NaCl/kg $CaCO_3$; the optimum concentration of the brine solution is about 10 percent NaCl.

TABLE 4.3

Regeneration of a Cation Exchange Resin

Regenerant level (lb NaCl/ft³)	Softening capacity (kg CaCO₃/ft³)	Salt efficiency (lb NaCl/kg CaCO₃)
5	17.2	0.29
10	25.0	0.40
14	28.0	0.50
15	29.1	0.52
20	32.5	0.62
25	34.0	0.74

Other factors that influence the capacity of a cation exchanger include the rate of flow, the particle size of the exchange material, the concentration of hardness in the influent, and the incidence of foreign materials such as iron floc, turbidity, silt, oil, and microorganisms, all of which interfere mechanically with the exchange reaction and also cause excessive pressure drop. Cation exchangers should always be protected by filters.

One of the more efficient cation exchange resins is cross-linked sulfonated polystyrene, prepared by copolymerizing styrene and polyvinylbenzene, then sulfonating most of the benzene rings. The capacity of this resin, which is stable throughout the entire pH range and up to 250 F, varies from 25,000-33,000 gr of calcium carbonate per cubic foot, with an average of about 28,000 in a full-scale installation. Over a period of time the capacity of a softener may decrease as a result of attrition and loss of resin particles, but this is a small factor compared with inactivation by the other causes mentioned above.

If the average hardness of a water supply is known, it is possible to calculate the number of gallons of water that can be softened in a softener of given size and exchange capacity. Thus, if the hardness of the influent water is 50 ppm, the volume of the bed of cation exchange resin is 150 ft^3, and its capacity is 29,000 gr of $CaCO_3/\text{ft}^3$, the volume of water that can be softened is

$$29,000 \times 17.1 \times 150/50 = 1,500,000 \text{ gal}$$

In any water treating plant the softeners are operated in parallel, discharging into a common header. The effluent of each unit is checked for hardness at regular intervals, e.g., every two hours, or continuously if automatic analyzers are available, and when the hardness rises above some fixed value, perhaps 2-5 ppm $CaCO_3$, the softener is disconnected from the header and the resin is regenerated. Five steps comprise the procedure:

1. *Backwash.* Piping and valves are provided for forcing water upward through the resin and out of the top of the shell into some sort of reclamation tank, or else to the sewer. This realigns the bed, washes out fines, and removes any accumulation of insoluble foreign material that may have collected on the surface of the resin during the preceding service run. The rate of backwashing should be such that the resin bed expands 50-75 percent of its normal volume ($\approx 6 \text{ gal}/\text{ft}^2/\text{min}$), if the water is at 70 F. Warmer water, being less viscous, will have to be added at a faster rate to obtain the same lifting effect.

2. *Brining.* After backwashing, the bed is allowed to settle, then a 10 percent solution of sodium chloride is introduced and allowed to

flow through the resin bed at a rate of 1 gal/ft^3/min; sufficient solution is used so that 15 lb of NaCl per cubic foot of resin is applied. The brine solution is drained to the sewer. Large softeners, with surface area $\geqslant$ 50 ft^2 are usually equipped with a brine distribution system consisting of several nozzles to spray the regenerating solution over the entire surface of the resin bed.

3. *Slow Rinse.* In carrying out, a regeneration allowance must be made for the void volume of the resin. This is the space between the spherical particles of resin, which in the cation exchangers under consideration, amounts to 35–40 percent of the total volume of the bed. Thus, after all of the brine has been introduced into the unit the rinse rate should be held at 1 gal/ft^3/min until all of the brine has been flushed out of the unit. Treated, softened water is used for this and the following rinse.

4. *Fast Rinse.* The rinse rate is finally increased to 1.5 gal/ft^3/min, and continued until the hardness in the effluent is less than 1 ppm CaCO$_3$, and the concentration of chloride does not exceed that of the rinse water by more than 25 ppm, as NaCl. This usually requires about 60 gal of water per cubic foot of resin.

5. *Run.* The softener unit is reconnected to the softener outlet header and returned to service at a flow rate of 2 gal/ft^3/min.

If water with a total hardness of < 100 ppm CaCO$_3$ is processed in a cation exchanger, the hardness in the effluent is essentially zero. In addition, cation exchange is especially suited for removing noncarbonate hardness, because sodium chloride is less expensive than sodium carbonate. There are several advantages in combining cation exchange with hot lime softening: residual carbonate is reduced, so less carbon dioxide is released later in the system; the elimination of phosphate and sodium carbonate reduces the total solids in the effluent; the cost of treatment is reduced by replacing sodium carbonate with sodium chloride.

At this point it is of interest to compare the costs of softening water by the lime soda process previously cited and by cation exchange. It was found that 83.8 ppm of hydrated lime and 26.5 ppm of soda ash were required to soften the water. The prices (current in November, 1980) shown in Table 4.4 are used to calculate costs.

Lime: lb/1000 gal   = 83.8 × 1000 × 8.35/10^6 = 0.699

$/1000 gal   = 0.699 × 0.0420

= 0.0294

$$\text{Soda ash:}\quad \text{lb/1000 gal} = 26.5 \times 1000 \times 8.34/10^6 = 0.221$$

$$\$/1000 \text{ gal} = 0.221 \times 0.0560$$

$$= 0.0124$$

Total chemical cost: $0.0294 + 0.0124 = 0.0418 \ \$/1000$ gal

TABLE 4.4

Chemical Costs

Chemical	$/ton	$/lb
Lime, Ca(OH)$_2$	84.00	0.0420
Soda ash, Na$_2$CO$_3$	112.00	0.0560
Salt, NaCl	22.80	0.0114

Cation exchange: Assume capacity of 29,000 gr $CaCO_3/ft^3$, a regenerant level of 15 lb $NaCl/ft^3$ resin, and a total hardness of 103 ppm $CaCO_3$.

$$\$/1000 \text{ gal} = \frac{15 \text{ lb/ft}^3 \times 0.0114 \ \$/\text{lb} \times 103 \text{ ppm } CaCO_3 \times 1000}{29,000 \text{ gr } CaCO_3/\text{ft}^3 \times 17.1 \text{ ppm/gr}}$$

$$= 0.0355$$

In Table 4.5 water analyses are given showing typical results obtained by the various methods of softening water that have been discussed. The values of hardness and alkalinity are in terms of equivalent ppm $CaCO_3$; the concentrations of the other ions are in parts per million of the individual ion.

Note that the total solids in any hard water are increased by sodium cation exchange. This is because the equivalent weight of sodium, 23, is larger than that of either calcium, 20, or magnesium, 12. Also, it is important to point out here that other cations (in particular, aluminum, ferrous, and ferric ions) are also retained by cation exchange resins. The protection of boilers against scales containing iron or aluminum such as $Na_2O \cdot Fe_2O_3 \cdot 4SiO_2$ (acmite) and $Na_2O \cdot Al_2O_3 \cdot 4SiO_2 \cdot 2H_2O$ (analcite) is an added benefit in installations using cation exchange.

TABLE 4.5

Comparison of Different Methods of Softening

Component	Untreated water	Cold lime-soda	Hot lime-soda	Hot lime-soda + PO$_4$	Cold lime and cation exchange
TH	103	35	20	2	1.0
CaH	68	22	15	1	0.5
MgH	35	13	5	1	0.5
HCO$_3^-$	78	0	0	0	0.0
CO$_3^{--}$	0	35	35	45	25.0
OH$^-$	0	5	5	5	5.0
Na$^+$	17	31	38	57	42.0
PO$_4^{---}$	0	0	0	8	0.0
Cl$^-$	12	12	12	12	12.0
SO$_4^{--}$	43	43	43	43	43.0

4.3 SILICA REMOVAL

The tendency of silica to form scales in boiler tubes requires that its concentration in feed water for boilers be reduced as much as possible. At pressures above 500 psi, silica selectively steam distills[5] forming hard, glassy, or amorphous deposits on the inside of superheater tubes and on turbine blades.[6] The efficiency of a turbine is drastically impaired by these scales. In Table 4.6 are listed maximum values for the concentration of silica in boiler water that ensure that no more than 0.02 ppm of silica will contaminate the steam from industrial boilers of modern conservative design. It is seen that permissible limits drop sharply, in fact, almost logarithmically, with increasing pressure.

a. Coagulation with Chemicals

The concentration of silica in moderately high-pressure steam generators (500-900 psi) can be conveniently controlled by modifying the methods of lime softening the make up water already described. If the cold lime softening process is used, ferric sulfate[7] can be added to form a floc of hydrous ferric oxide, which adsorbs silica. In unheated water this treat-

TABLE 4.6

Permissible Concentrations of Silica in Boilers

Maximum pressure (psi)	Maximum silica (ppm)
500	40.0
600	35.0
750	30.0
900	20.0
1000	10.0
1500	3.0
2000	1.0
2500	0.5

ment is most successful when carried out in the pH range of 9.0-9.5; a residual of 2-3 ppm of silica is normally obtained.

If one of the hot processes is used for softening, magnesium hydroxide is effective for reducing the concentration of silica; a residual of 0.5-1.0 ppm SiO_2 can be realized at the optimum pH value of 10.2. Better results are obtained if magnesium is added as the ion and precipitated in the presence of silicate. This is because magnesium silicate is precipitated directly, a much more rapid reaction than that of silicate with insoluble magnesium hydroxide. The latter reaction is aided by using "activated" magnesium oxide, which is said to hydrate more rapidly than does that obtained from dolomitic lime, for example. The ratio of added MgO to SiO_2 in the water to be treated should be three to one. The time of contact between water and floc must be fairly long, however, and for this reason processes using sludge blankets are particularly suited for the adsorption of silica by magnesium hydroxide. In a treatment with a sludge blanket the water is intimately in contact with active floc for sufficient time for equilibrium to be established between the coagulant and the silica dissolved in the water. For best results the alkalinity is controlled so that the differential, (2P − M), is 12-18 ppm $CaCO_3$. Note also that if ferric sulfate is used as the coagulant extra lime must be added to maintain the proper free alkalinity.

C. H. Spaulding[8,9] is the inventor of an ingenious treater called the Spaulding Precipitator, consisting of a truncated conical vessel containing a mixing chamber in its center shaped like an inverted funnel. The treating chemicals, fed as slurries, are introduced with the water to be treated into the center chamber. Mixing is accomplished by a large mechanical agitator consisting of several paddles rotating at 1-3 rpm. Water flows downward in the mixing chamber, then upon passing out of the inverted funnel, reverses direction, and flows upward, the vertical velocity continuously decreasing because of the conical shape of the vessel. At a certain level there is a separation of water from the floc producing, in ideal operation, a sharply defined sludge level. The level is held constant by continuously bleeding off spent sludge from a special collection chamber at the bottom of the unit. As in the lime softening processes, recirculating a small volume of active sludge greatly improves floc formation.

Some quantitative relationships among the operating variables of an actual Spaulding Precipitator must be derived in order to calculate the retention time of water in the unit, the length of time the water is in contact with the sludge blanket, and the rate at which the water rises when at any particular height above the bottom of the unit. It is assumed that the rate of water through the unit is 600,000 lb/h, or 160 ft^3/min. The top diameter of the vessel is 58 ft, the bottom diameter is 36 ft, and the height is 14 ft. The total volume of the vessel is

$$V = \pi h(r_1{}^2 + r_2{}^2 + r_1 r_2)/3$$

$$= 14\pi(18^2 + 29^2 + 18 \times 29)$$

$$= 24{,}733 \ ft^3$$

$$\cong 185{,}000 \ gal$$

Fig. 4.2 shows that the precipitator is an inverted frustum of a right circular cone. Using the methods of analytic geometry, first the height of the imaginary cone formed by extending $P_1 P_2$ to P_0 is calculated. The dimensions of the actual precipitator are shown in Fig. 4.3. To find the total height, s, y_0 is calculated, and the height of the precipitator (14 ft) is added to its absolute value.

$$\text{Slope}, m, \text{of } P_0 P_1 P_2 = (y_2 - y_1)/(x_2 - x_1) = (14 - 0)/(29 - 18)$$

$$= 14/11$$

The equation of $P_o P_1 P_2$ is:

$$(y - y_1) = m(x - x_1)$$
$$y = 14(x - 18)/11$$
$$11y = 14x - 252$$

When $x = 0$, y_o is the apex of the imaginary cone.

$$y_o = -252/11$$
$$= -22.9$$

Therefore, the total height, s, of the imaginary cone is

$$14 + 22.9 = 36.9 \text{ ft}$$

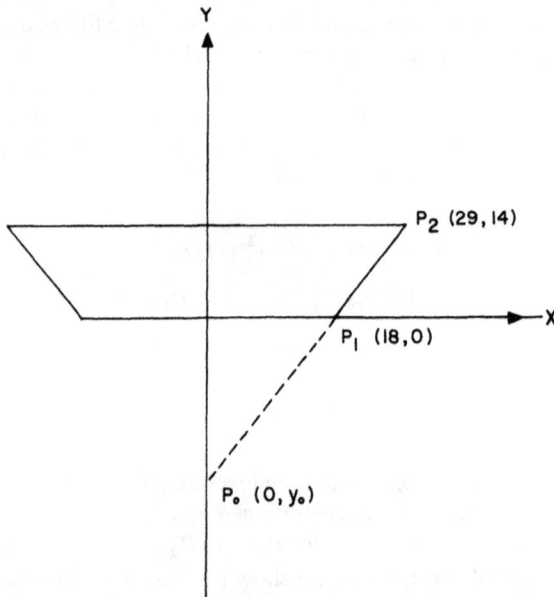

Fig. 4.2 Calculation of the total height of the imaginary cone.

Now the rate at which water is rising at any value of h, where h is the height of an imaginary circular water surface above the bottom of the precipitator, and r is the radius of that circular surface at height, h can be calculated.

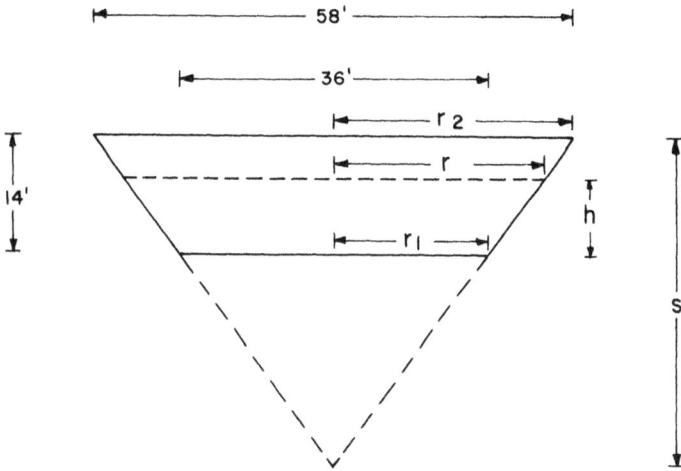

Fig. 4.3 Diagram of the Spaulding Precipitator

If r is expressed in terms of h by noting in Fig. 4.3 that

$$r : r_2 = (h + s - 14) : s$$

Therefore,

$$r = (h + s - 14)r_2/s$$

or

$$r = (h + 36.9 - 14)29/36.9$$

The volume of water that would fill the entire imaginary cone to height, h, is

$$V = \pi r^2 (h + 22.9)/3$$
$$= \pi (29/36.9)^2 (h + 22.9)^3/3$$
$$= 0.647(h + 22.9)^3$$

To find the rate of change of V with respect to time, t, the last equation is differentiated.

$$dV/dt = 1.94(h + 22.9)^2 \ (dh/dt)$$

$$= 160 \ \text{ft}^3/\text{min (the given water flow rate)}$$

Therefore,

$$dh/dt = 160/1.94(h + 22.9)^2$$

$$= 82.5/(h + 22.9)^2$$

From this last expression, the rates in Table 4.7 at various heights are calculated.

TABLE 4.7

Rate of Water Rise in Spaulding Precipitator

Height (ft)	Rate of rise (ft/min)
0	0.157
3	0.123
6	0.099
9	0.081
12	0.068
14	0.061

To calculate the time that the water is in the unit the variables in the differential equation derived above are separated, and solved for by integration.

$$dh/dt = 82.5/(h + 22.9)^2$$

$$\int_0^h (h + 22.9)^2 \, dh = 82.5 \int_0^t dt$$

$$(h + 22.9)^3/3 \Big]_0^h = 82.5t \Big]_0^t$$

$$[(h + 22.9)^3 - (22.9)^3]/3 = 82.5t$$

$$[(h + 22.9)^3/3 - 4003] = 82.5t$$

Therefore, the time required for the water to rise from the bottom of the precipitator to the collection trough at the top is

$$t = [(14 + 22.9)^3/3 - 4003]/82.5$$
$$= 154 \text{ min}$$

If the top of the sludge blanket is 6 ft below the upper surface of the water in the precipitator, the length of time the rising water is in contact with the sludge blanket is

$$t = [(8 + 22.9)^3/3 - 4003]/82.5$$
$$= 70.7 \text{ min}$$

In Table 4.8 several values are tabulated that are pertinent to the operation of a Spaulding Precipitator. This particular unit is operated cold, and the principal concern is reducing the concentration of silica; the effluent is subsequently filtered and then softened by cation exchange. More specific operating details are found in Chapter 6, pp. 147-149. The samples in Table 4.8 were taken at a time when the treatment was as follows: lime 24 ppm, magnesia 36 ppm, ferric sulfate 5.4 ppm. Comparing the P and M values of the filtered treater effluent shows that the free alkalinity, $(2P - M)$ is too low, which accounts for the high concentration of magnesium hardness in the effluent. As mentioned before, this value should fall in the range of 12-18 ppm $CaCO_3$.

TABLE 4.8

Operation of a Spaulding Precipitator

Sample	Total hardness	Calcium hardness	Magnesium hardness	P	M	SiO$_2$ (ppm)
Untreated water	58	40	18	1	47	8.8
Treater effluent (filtered)	58	23	35	24	45	2.1
Treater effluent (acidified)	84	36	48	--	--	--

Acidifying an unfiltered sample of effluent dissolves the suspended floc. By comparing the distribution of hardness in filtered and in acidified portions of the effluent from the precipitator the percentage of carryover, i.e., the amount of floc that must subsequently be removed by filtering, and the approximate composition of the sludge in the treater can be calculated.

Using the results derived in Section 4.2a for calculating the dosage of lime required for cold lime softening, and noting that the calcium hardness is less than the M-alkalinity, the following can be calculated:

$$\text{lime required} = 0.742(M + MgH)$$

$$= 0.742(47 + 18)$$

$$\cong 48 \text{ ppm Ca(OH)}_2$$

Ferric sulfate is added as required to maintain the value of $(2P - M)$ in the range of 12-18 ppm $CaCO_3$; magnesium oxide is added to yield the lowest possible concentration of silica in the effluent. Using the values for the distribution of hardness in the samples of effluent from the treater, the following can be calculated:

Increase in total hardness: None
Increase in magnesium hardness: $(17/18) \times 100 = 94.5$ percent
Calcium precipitated: $(17/40) \times 100 = 42.5$ percent
Magnesium in floc: $(48 - 35) \times 100/(84 - 58) = 50$ percent
Calcium in floc: $(36 - 23) \times 100/(84 - 58) = 50$ percent
Carryover: $(84 - 58)/10^4 = 0.0026$ percent

Carryover can be reduced to less than 10 ppm by adding 1-2 ppm of a nonionic polymer with the ferric sulfate. Hydroxyethylcellulose,[10] at a concentration of 1-2 ppm, has also been recommended for toughening floc. It is prepared as a 5 percent slurry in a 5 percent solution of sodium hydroxide. The slurry can be added to the lime slurry tank, or else proportioned directly into the reaction vessel, after diluting 5 : 1. As hydroxyethylcellulose is precipitated in acidic solutions it cannot be added to the ferric sulfate mixer.

b. Demineralization

The stringent limitation on the concentration of dissolved solids in steam generators operated at 1000 psi and above, requires that exceptionally pure feed water be used. Accordingly, the make up water for these

TABLE 4.9

Variables in the Operation of a Sludge Blanket Treater*

Variable	Function	Effect of deficiency	Effect of excess	Effect of increase
Lime (CaO)	Lime softening. Floc formation. Source of alkalinity.	Excess hardness in effluent. Turbidity in effluent. Low differential.	Excess hardness in effluent. High differential.	Differential increases.
Magnesia (MgO)	Silica removal. Floc formation.	Poor silica removal. High density of floc.	Turbidity in effluent.	None immediate. Long-term improvement in removal of silica.
Ferric sulfate (Fe$_2$(SO$_4$)$_3$)	Silica removal. Binder for floc. Bed level control. Differential control.	High differential. Turbidity in effluent. Low floc density.	Iron in effluent. Low differential. Poor silica removal.	Differential decreases. Bed level rises.
Agitator speed	Mixing.	High sludge density. Low bed level.	Low sludge density. High bed level.	Bed level rises.
Desludging rate	Removal of spent sludge. Bed level control.	Inactive sludge. High bed level.	Low density sludge. Variable bed level.	Bed level falls.
Differential (2P − M)	Alkalinity control.	Turbid effluent.	Poor silica removal.	Silica increases in effluent. pH of effluent rises.

* This table summarizes the effects of several variables that are important in the operation of a sludge blanket treater for removing silica by the use of active floc.

boilers more often than not is demineralized. This method of purifying water is especially popular in Europe where practically all power stations with high pressure boilers use demineralized water for make up. Ordinarily, the feed water for these boilers is predominantly condensate, with only a small percentage of treated make up.

Earlier in this chapter sodium cation exchange using sulfonated polystyrene resin was described as a process for softening water. When this resin is transformed to the hydrogen form, and followed by a strong-base anion exchanger such as quaternary ammonaited polystyrene in the hydroxide form, an effluent is produced containing very small concentrations of soluble ions. Typical exchange reactions are represented by Eqs. (4-10) and (4-11), in which R represents the cross-linked polystyrene resin matrix.

Cation exchange:

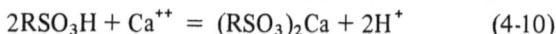

$$2RSO_3H + Ca^{++} = (RSO_3)_2Ca + 2H^+ \qquad (4\text{-}10)$$

Anion exchange:

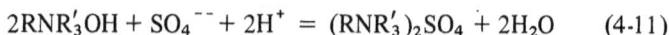

$$2RNR_3'OH + SO_4^{--} + 2H^+ = (RNR_3')_2SO_4 + 2H_2O \qquad (4\text{-}11)$$

Thus, cations and anions, in equivalent amounts, enter the cation exchanger, and hydrogen ions equivalent to the cations in the influent emerge from the cation exchanger along with the anions. The latter are next exchanged for hydroxide ions in the anion exchange resin, and these are neutralized to form water by the equivalent number of hydrogen ions from the cation exchange reaction. When demineralized water containing high concentrations of bicarbonates, it is advisable to install a degasifier, or spray-type carbon dioxide extractor, between the cation and anion exchangers to eliminate carbon dioxide formed in the cation exchange.

$$RSO_3H + Na^+ + HCO_3^- = RSO_3Na + H_2O + CO_2 \qquad (4\text{-}12)$$

If carbon dioxide enters the anion exchanger, it reacts as shown in Eq. (4-13).

$$RNR_3'OH + CO_2 = RNR_3'HCO_3 \qquad (4\text{-}13)$$

Also, if large amounts of carbon dioxide are released in the cation exchanger, gas pockets form in the bed that mechanically interfere with the exchange reaction.

Strong-base anion exchange resins adsorb silicate ion while simultaneously neutralizing meta-monosilicic acid, H_2SiO_3, which is the predominant species in natural water. In acidic solution this compound is prone to polymerize, first to orthosilicic acid, H_4SiO_4, then to a colloidal solution, or hydrogel, having the general formula, $mSiO_2 \cdot nH_2O$. This polymerization is minimized by using warm sodium hydroxide to regenerate the anion resin. Colloidal silica and polymeric silicic acids that enter reservoirs during heavy rains pass through anion resins unchanged, and so into the boiler. Conventional demineralizers, as just described, yield an effluent containing 1-10 ppm of dissolved solids, with an average concentration of silica of about 0.1 ppm.

The strong-base anion exchange resins used for complete demineralizing and removing silica, contain two functional groups: a quaternary ammonium group (80-85 percent), and a tertiary amine group (15-20 percent), of which only the former can exchange silica. The total capacity of the resin (about 30,000 gr $CaCO_3/ft^3$ when new) is the sum of the capacities of the two functional groups, but only the quaternary ammonium, or salt-splitting, group is useful for demineralizing water for boiler feed make up. Methods for evaluating ion exchange resins are given in Chapter 8.

Anion exchange resins degrade rapidly above 140 F, are easily contaminated by organic material, and are oxidized and inactivated by chlorine and other oxidizing agents. Iron combined with organic compounds such as tannins and humic acids is often found in anion exchange resins. Degradation products of cation exchange resins, as well as many organic acids, are irreversibly adsorbed on the anion exchangers, decreasing their exchange capacities—often to unacceptably low values. This can be avoided by interposing an activated carbon treater, or a precoat filter containing activated carbon and diatomaceous earth, between the two exchanger vessels to remove the organic materials. The most complete removal of organic matter is realized if the pH of the water can be adjusted to the iso-electric point of the organic species to be removed. Inorganic particles and some colloids are also retained. Residual chlorine, for example, in municipal water supplies, destroys the cross-linking of the resin copolymers, increasing their capacity for holding moisture and also attacks the resin directly by oxidizing the quaternary ammonium group to the inactive tertiary amine. This latter reaction is catalyzed by elements of the first transition series, notably iron, cobalt, nickel, and copper.

Except for a few essential differences, the procedure for regenerating the cation and anion resins is similar to that described on pp. 48–49. When the anion resin is saturated silica appears first in the effluent, followed by chloride and a decrease in pH. Both the cation and anion exchangers are backwashed to eliminate fines and miscellaneous fouling, and to reclassify the beds. The cation resin is regenerated starting with 2 percent sulfuric acid at 2 gpm/ft^3, followed by 4 percent acid at 1.5 gpm/ft^3, and finally 6 percent acid at 1 gpm/ft^3; a total of 15 lb of 66 Bé H$_2$SO$_4$/ft^3 resin is used. The acid treatment is carried out this way to minimize the precipitation of calcium sulfate within the resin beads, which alternately shrinking and swelling around the crystals, would eventually fracture. A special porous anion exchange resin is best for treating water containing large organic molecules.

Anion exchange resins are regenerated by passing warm (120 F) 4 percent sodium hydroxide through the bed at 0.25 gpm/ft^3 for one bed volume, then at 1.5 gpm/ft^3, using a total of 8-12 lb NaOH/ft^3 resin, depending upon the completeness of silica removal required. Rayon grade or mercury cell grade sodium hydroxide should be used for the regeneration, as these grades contain very small concentrations of chloride and chlorate ions. The former limits the over-all capacity, and the latter can act as an oxidizing agent in some circumstances, although its reaction rate as such is very slow.

For optimum removal of silica the bed should be preheated by passing warm (not over 125 F) condensate through the resin before returning the unit to service. The normal exchange rate for demineralizing make up water is 2-5 gpm/ft^3 of resin. In general, at least 3 ft of bed depth should be used, but note that excessive pressure drop breaks the resin beads.

Should it be necessary to reduce the concentration of silica to less than 0.05 ppm, it can be achieved by adding a mixed bed deionizing unit after the anion exchanger, but this is usually unnecessary except for preparing make up for supercritical boilers. Balthazar[11] reports that up-flow regeneration of demineralizers, which is widely practiced in Europe, greatly improves the quality of water produced and appreciably reduces the consumption of regenerating chemicals and rinse water. A special procedure is used including the following steps.

1. The exhausting run is discontinued before leakage commences.
2. Decationized water is used to dilute the sulfuric acid and for the up-flow rinse of the cation exchanger.
3. Demineralized water is used to dilute the sodium hydroxide and for the up-flow rinse of the anion exchanger.
4. The slow up-flow rinses are prolonged so that during the fast down-

flow rinse no ions are left in the resin that are supposed to be eliminated during the exhausting run.

Multistage deionizing units, required to "polish" condensate used in feed
water for supercritical boilers, are discussed in Chapter 7. Table 4.10 shows
the concentrations of residual silica produced by the various methods for
removing silica discussed in this chapter.

TABLE 4.10

Silica Removal by Various Methods

Process	Residual SiO_2 (ppm)
Cold lime softening	2–3
Hot lime softening–Cation exchange	0.5–1.0
Conventional demineralizing	0.05–0.2
Conventional demineralizing plus mixed bed	0.01–0.02
Demineralizing with up-flow regeneration	<0.01

REFERENCES

(1) Black, A. P. 1960. Basic mechanisms of coagulation. *J. Amer. Water Works Assoc.* 52:492.

(2) Wood, F. O. 1972. Selecting a softening process. *J. Amer. Water Works Assoc.* 64:820.

(3) Fluid Process Chemicals Dept. 1970. *Amberlite Ion Exchange Resins.* Philadelphia: Rohm and Haas.

(4) Nachod, F. C. and Schubert, J. 1956. Ion exchange technology. New York: Academic Press.

(5) Coulter, E. F., Pirsh, E. A., and Wagner, E. J., Jr. 1956. Selective silica carry-over in steam. *Trans. Amer. Soc. Mech. Engrs.* 78:869.

(6) Straub, F. G. and Grabowski, H. A. 1945. Silica deposition in steam turbines. *Trans. Amer. Soc. Mech. Engrs.* 67:309.

(7) Schwartz, M. C. 1938. The removal of silica from water for boiler feed purposes: the ferric sulfate and hydrous ferric oxide process. *J. Amer. Water Works Assoc.* 30:659.

(8) Spaulding, C. H. 1937. Conditioning of water softening precipitates. *J. Amer. Water Works Assoc.* 29:1697.

(9) Spaulding, C. H. 1938. Some new practices in water softening. *Water Works and Sewerage* 85:153.

(10) Kemmer, F. M., Hauser, C. B., Steeper, R. E., and Dickinson, B. W. 1952. A new coagulation aid. *Water and Sewage Works* 99:16.

(11) Balthazar, J. 1967. Some aspects of water treatment in european power stations. *Proc. Amer. Power Conf.* 29:789.

Chapter 5.

Internal Chemical Treatments

There are a number of treatments that are made within the boiler to minimize the adverse effects of small concentrations of components that remain in the feed water after the preliminary operations discussed in Chapters 3 and 4 that are now considered.

5.1 PRECIPITATION OF CALCIUM AND MAGNESIUM

a. Phosphate

Regardless of the method used for softening, a certain amount of residual hardness is always present in treated water. One of the most widely used methods for preventing alkaline earth scales in boilers up to 2000 psi is to add phosphate, which in alkaline solution precipitates calcium and magnesium as a soft dispersed sludge. A number of phosphate salts are used for water conditioning, the choice depending to a great extent upon considerations of alkalinity. In Table 5.1 five commonly used phosphates are listed with their formulas, solubilities, pH of a 100-ppm solution, and their reaction in boiler water.

The first three salts are o-phosphates, which form insoluble basic salts with calcium and magnesium ions. Monobasic sodium phosphate is acidic, and if added to a boiler it reduces the concentration of hydroxyl ion in the water. In the pH range of 8-11 the predominant o-phosphate ion is HPO_4^{--}, which means that adding disodium phosphate to alkaline boiler water does not appreciably affect the alkalinity. Trisodium phosphate is used in systems in which the feed water contains little or no bicarbonate—demineralized or evaporated make up, for example.

If o-phosphate is added to feed water calcium phosphate precipitates in feed water lines and preheaters causing plugging. The best method for introducing phosphate is by injecting a solution of disodium phosphate directly into the steam drum from a pressure pot or with a positive displacement pump actuated by a timer. The chemical feed should enter close to the downcomer tubes to avoid interference with, or short-circuiting through, the continuous blowdown line.

TABLE 5.1

Chemical Properties of Various Phosphate Salts

Chemical name	Formula	Solubility (g/l @ 20 C)	pH (100 ppm)	Reaction in boiler
Monobasic sodium phosphate	$NaH_2PO_4 \cdot H_2O$	110.3	5.1	Acidic
Dibasic sodium phosphate	$Na_2HPO_4 \cdot 12H_2O$	12.0	8.6	Neutral
Tribasic sodium phosphate	$Na_3PO_4 \cdot 12H_2O$	25.8	10.3	Alkaline
Sodium tripoly-phosphate	$Na_5P_3O_{10}$	64.0	9.7	Acidic
Sodium hexameta-phosphate	$(NaPO_3)_6$	973.0	5.8	Acidic

If a chemical injection system is not provided polyphosphates may be used; they are hydrolyzed rapidly in the boiler to o-phosphate by the hot alkaline water. The following equations represent the hydrolysis, or reversion, of the two polyphosphates listed in Table 5.1.

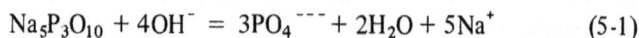

$$Na_5P_3O_{10} + 4OH^- = 3PO_4^{---} + 2H_2O + 5Na^+ \qquad (5\text{-}1)$$

$$(NaPO_3)_6 + 12OH^- = 6PO_4^{---} + 6H_2O + 6Na^+ \qquad (5\text{-}2)$$

These equations show that the free alkalinity of the boiler water is reduced when polyphosphates are used. These salts are particularly valuable when applying the coordinated phosphate-pH method of alkalinity control to feed water containing an appreciable concentration of carbonates, as described further on in Section 5.4b.

The reversion of polyphosphates has been discussed by Green.[1] On account of hydrolysis they may not be usable if feed water temperature is high, the feed water line is long, or the retention time in the preboiler system is great. The rate of hydrolysis is slow in cold water, but increases rapidly with increasing temperature, acidity, or alkalinity. The presence of a precipitable ion, such as calcium, also favors the hydrolysis by shifting the equilibria represented by Eqs. (5-1) and (5-2).

If polyphosphates remain in the molecularly dehydrated state they form stable complexes with calcium, magnesium, aluminum, iron, and other cations, and so do not precipitate until they enter the boiler and hydrolysis occurs. In aqueous solution at pH 7.5, for instance, the predominant ion of tripolyphosphate is $HP_3O_{10}^{----}$. This ion forms a soluble ion-pair complex with calcium ion, $CaHP_3O_{10}^{--}$, the dissociation constant of which is 1.04×10^{-4}. Jenkins, et al.[2] point out that at pH 7.5 about 90 percent of tripolyphosphate exists in the solution as $CaHP_3O_{10}^{-2}$. Thus, the stability of this complex ion prevents the precipitation of insoluble calcium salts in boiler feed lines. Magnesium forms a slightly stronger complex ion; iron and aluminum form much stronger complexes.

The precipitate formed by calcium and o-phosphate is usually represented as the tribasic salt, $Ca_3(PO_4)_2$. At the high alkalinities of boiler waters, however, this salt is prone to form supersaturated solutions, so the nucleation and crystal growth rates are slow. In boilers the precipitate invariably obtained is hydroxyapatite, $Ca_5(PO_4)_3OH$, which forms a soft nonadherent sludge that is less likely to coat tubes than either calcium carbonate or basic calcium carbonate $[Ca(OH)_2 \cdot CaCO_3]$.

Basic phosphates of magnesium tend to stick to hot metal and do not respond well to dispersants. Also, precipitates of both calcium and magnesium formed with phosphate when the pH of the water is less than 9.6 tend to be sticky. In the absence of dispersants, the basic phosphates may adhere to boiler tubes; this can be minimized by maintaining a high pH, i.e., 11-12.

Table 5.2 lists suitable phosphate residuals in boilers at various pressures up to 2000 psi, together with maximum permissible hardness in the corresponding feed waters.

It is not practicable to use phosphate when the total hardness of the feed water is in excess of 15-20 ppm $CaCO_3$ because so much sludge is formed that excessive blowdown is required. In lower pressure boilers sodium carbonate can be used to precipitate the hardness in the feed water. Care should be taken not to overfeed sodium phosphates, as high concentrations cause foaming and the separation of sodium phosphate muds in boiler drums and in water-wall headers.

TABLE 5.2

Suggested Ranges for Phosphate

Maximum pressure (psi)	Phosphate in boiler water (ppm)	Maximum hardness in feed water (ppm $CaCO_3$)
100	NR	75.00
200	40–50	20.00
300	30–40	2.00
500	25–30	2.00
600	20–25	0.20
750	15–20	0.10
900	10–15	0.05
1000	5–10	0.05
1500	3–6	0.00
2000	1–3	0.00

NR = not recommended

b. Sodium Carbonate

In using sodium carbonate the emphasis is on preventing the formation of anhydrite, $CaSO_4$, which crystallizes *in situ* as a hard adherent scale directly on hot surfaces. In addition to calcium sulfate, calcium meta-monosilicate, $CaSiO_3$, and hydrated magnesium orthodisilicate, $Mg_3Si_2O_7 \cdot 2H_2O$, also deposit this way from boiler water. That is, the crystals are not present as individual particles in the water previous to their appearance on the hot surface. Calcium carbonate and magnesium hydroxide are often found in these scales, but only as incidental inclusions of loose crystals in the interstices of the matrix of one or more of the other three scales.

Hall, et al.[3] have shown that to prevent the formation of anhydrite scale by carbonate it is necessary that the following relationship be satisfied:

$$(CO_3^{--}) > \left(\frac{K_{sp} \ CaCO_3}{K_{sp} \ CaSO_4} \right)_T (SO_4^{--}) \tag{5-3}$$

For example, at 365 F and 150 psi the formation of $CaSO_4$ is prevented by making sure that

$$\text{ppm } CO_3^{--} > 0.0883 \times \text{ppm } SO_4^{--} \tag{5-4}$$

Calcium carbonate formed in this treatment produces a thin, porous, flaky layer on evaporative surfaces so dispersants (conditioners) are also required.

The use of sodium carbonate to prevent scaling is limited to boilers operated at less than 200 psi. At greater pressures high caustic alkalinity develops from the decomposition of excess carbonate, making it impossible to maintain the proper ratio of carbonate to sulfate without excessive blowdown. The treatment is most successful with feed waters in which the calcium hardness is less than the M-alkalinity.

5.2 CHELATION OF CALCIUM AND MAGNESIUM

The enthusiasm with which the introduction of chelating agents was greeted has moderated to some extent as it became evident that their application was not a panacea in the practice of boiler water treatment. Nevertheless, under careful control and proper operating conditions they are eminently satisfactory for controlling scaling by alkaline earth metals. Chelating agents keep boilers remarkably clean by forming soluble complexes with calcium and magnesium, rather than the insoluble salts characteristic of phosphate or soda ash treatments.

a. Chemistry of Chelation

Bell[4] has discussed chelation chemistry as it relates to water treatment and chemical cleaning. The two chelants most generally used in water treatment are the sodium salts of ethylenedinitrolotetraacetic acid (EDTA) and nitrilotriacetic acid (NTA). EDTA forms soluble quadridentate chelonates with calcium and magnesium ions.

In chelonates the cation is bonded at a minimum of two sites on the ligand to form a ring structure, which is inherently more stable than ordinary coordination complexes. The arrows in the structural formula indicate coordination covalent bonds. The stability of chelonates is measured by the magnitude of their formation constants–the larger the positive value of K_f, the greater the tendency for the chelonate to form. If M^{++} is the cation and A^{--} is the anion of the chelant

$$K_f = (MA)/(M^{++})(A^{--}) \qquad (5\text{-}5)$$

In Table 5.3 the values given by Meites[5] of the logarithms of K_f for a number of complexes of EDTA and NTA with various cations are shown.

TABLE 5.3

Formation Constants of Selected Chelonates with EDTA and NTA

Cations	EDTA	NTA
Mg^{++}	8.7	7.0
Ca^{++}	10.7	8.2
Fe^{++}	14.4	8.8
Fe^{+++}	25.1	15.9
Cu^{++}	18.8	12.7
Zn^{++}	16.6	10.5
Co^{++}	16.3	10.6
Al^{+++}	16.1	—

In interpreting the formation constants Bell[4] points out that it is necessary to take into account competitions among cations and anions for each other. The most important of these are the effects of pH, solubility, and stability of the chelonates. Thus, in acidic solution the affinity of the chelant for hydrogen ions weakens its chelating potential. Similarly, and despite claims to the contrary, the extreme insolubility of ferric hydroxide ($K_{sp} = 6 \times 10^{-38}$) precludes the possibility of forming a ferric chelonate in alkaline solution.

Bell[4] has formulated an exchange reaction to predict whether a chelant can dissolve a slightly soluble compound.

$$MA + B = MB + A \qquad (5\text{-}6)$$

In words: chelonate plus precipitant equals insoluble salt plus chelant. The exchange constant for this reaction is calculated from the formation constant of the chelonate and the solubility product of the insoluble compound.

$$K_e = 1/K_f \times K_{sp} \qquad (5\text{-}7)$$

If $K_e < 1$, i.e., if $(K_f \times K_{sp}) > 1$, then the chelant will dissolve the insoluble compound or prevent its precipitation. For example, if the insoluble compound is $CaCO_3$, $K_{sp} = 4.8 \times 10^{-9}$, and the chelant is EDTA, which has a formation constant with calcium, $K_f = 5.0 \times 10^{10}$, then

$$K_e = 1/5.0 \times 10^{10} \times 4.8 \times 10^{-9}$$
$$= 1/240$$

indicating that EDTA will dissolve an existing scale of calcium carbonate.

b. Application of Chelants to Boiler Water Treatment

The use of EDTA for preventing scales in boilers was first tested by Edwards and Rozas[6] in a 150-psi water-tube boiler. Subsequently, the treatment was applied to boilers operating at 1375 psi.[7] In the latter application EDTA showed no tendency to thermally decompose and it was effective in keeping calcium and magnesium in solution without the necessity for dispersants. According to Jacklin,[8] NTA is more efficient than EDTA at pressures up to 800 psi, but at 1200 psi they are equal. At 1500 psi EDTA is superior. More recently it has been averred[9] that mixtures of the two chelants are more effective than either used alone.

Upon initiating a chelant treatment suspended iron oxide appears in the blowdown water, but it disappears eventually. Also, upon replacing the coordinated phosphate-pH system with EDTA there is a considerable increase in free caustic alkalinity $(2P - M)$. Neither EDTA, nor NTA dissolve deposits of iron oxide, nor do they prevent aluminum ion in the feed water from forming analcite scale.

Lux[10] cautions that close control is necessary when using chelants, to avoid excessive corrosion, particularly in the presence of dissolved oxygen. Chelants should be added to the feed water rather than being injected into the steam drum; stainless steel equipment is essential for handling concentrated solutions. Some engineers recommend maintaining a 10-15 ppm residual of phosphate in case of a deficiency of chelant to ensure that no hard scale is deposited, but the rationale of this practice is questionable.

Catalyzed sodium sulfite must not be mixed with EDTA or NTA solutions, as the cobalt catalyst is inactivated making the reduction of dissolved oxygen slow and increasing the rate of corrosion. Sodium sulfite should be added as far ahead as possible of the chelant, preferably to the storage section of the deaerator, so that oxygen is reduced before the chelant enters the water. Also, when boilers under chelant treatment are shut down Lux[10] recommends nitrogen-blanketing to avoid attack when the boiler cools and air is drawn in.

Schantz[11] reports that boilers under chelant treatment show less adverse effects of oil than do those under phosphate treatment. He also found NTA effective up to 800 psi; EDTA was satisfactory at 1500 psi. There is some indication that NTA is more corrosive than EDTA, but the formation constants in Table 5.3 show that EDTA forms the more stable chelonates of the two. Furthermore, NTA is three to four times more expensive than EDTA, which itself costs two to three times as much as phosphate treatment. If the total hardness in feed water exceeds 2 ppm $CaCO_3$ a chelant program cannot be justified economically. As 416 g of Na_4EDTA sequesters 100 g of $CaCO_3$, the dosage required is 4.2 ppm EDTA/ppm total hardness in the feed water. Lux[10] recommends that a residual of EDTA not less than 5-10 ppm be maintained in the boiler water. Therefore, a 5-10 percent excess should be added to the feed water to obtain a free residual in the boiler water. The daily requirement of EDTA can be calculated by the following formula:

$$\#EDTA/day = (\#EDTA/\#CaCO_3) \times TH (ppm\ CaCO_3) \times$$

$$\times\ steam\ rate\ (\#/h) \times 24\ (h/day)/10^6$$

$$= 4.2 \times TH \times steam\ rate \times 24/10^6$$

In Chapter 8 a procedure is given for determining EDTA residual in boiler water. Meticulous care is essential in controlling a chelant program to avoid serious damage to boilers, but under proper control corrosion rates in the range of 0.1-0.4 mpy can be maintained.

5.3   INHIBITION OF CORROSION IN BOILERS

In common with most methods for inhibiting corrosion, the protection of boiler metal depends upon the formation and maintenance of a thin protective film on the surface of the metal. In boilers the protective film is magnetite formed by the reaction of iron with hot water according to Eq. (5-8).

$$3Fe + 4H_2O = Fe_3O_4 + H_2 \qquad (5\text{-}8)$$

As magnetite is the only significant product of corrosion at the temperatures of boilers,[12] it is of interest to examine the mechanism of its formation.

Many years ago Pfeil[13] elucidated this unusual mechanism in an elaborate study. He found that the coating of magnetite formed on hot surfaces in air is composed of three distinct layers showing a definite composition gradient between them. The layer in contact with the metal is porous, with 30-50 percent voids and 85 percent FeO and 15 percent Fe_2O_3. With alloy steels this layer is free of alloying metals. The next layer is found to be lightly etched with the pattern of the original surface of the metal; it consists of 70 percent FeO and 30 percent Fe_2O_3. The outer layer is black and glassy, with a smooth, velvety appearance; its composition is 20 percent FeO and 80 percent Fe_2O_3.

Pfeil hypothesized that at the iron-scale interface, iron is continually being dissolved as FeO in the scale forming an iron-rich solution. There is, thus, a countercurrent diffusion of iron outward and of oxygen inward, iron atoms leaving the core metal without disturbing the scale already formed. He supported this by showing that scale with higher oxygen content reacts with scale of lower oxygen content; that metallic iron is oxidized by oxygen-rich scale; and that iron does indeed pass outward through the layers of scale to the surface. If oxygen diffused inward and reacted with iron at the metal surface, the increase in volume would fracture the scale already formed. Therefore, iron must also diffuse outward. Oxygen for this reaction can, of course, be supplied by sources other than air, including carbon dioxide, water, and cupric oxide.

The ideal magnetite layer is a diffusion barrier of sufficient thickness and porosity to slow the rate of oxidation of the metal without impeding heat transfer significantly. For this reason, it should be dense rather than porous to avoid trapping boiler water, which would subsequently concentrate under the layer of scale. Douglas and Zyzes[14] conclude that in a boiler, Fe_3O_4 reaches a critical thickness beyond which it becomes porous and the corrosion rate becomes constant as a result of resistance by the film to the migration of atoms of iron. This film is most stable in the pH range of 11-12 and consequently all boilers are operated with alkaline water.

a. Control of Alkalinity

The determination of various types of alkalinity by titration with a standard solution of acid is a simple and important method for controlling the operation of boilers. By measuring the alkalinities indicated by phenolphthalein and methyl orange (P and M values) it is possible to calculate the approximate concentrations of hydroxide, carbonate, and bicarbonate in most water samples. These values are significant in controlling corrosion and preventing scaling and foaming in boilers. The reactions are:

P-alkalinity (pH 8.3)

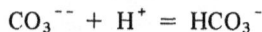

$$OH^- + H^+ = H_2O$$
$$CO_3^{--} + H^+ = HCO_3^-$$

M-alkalinity (pH 3.9)

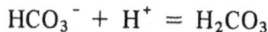

$$OH^- + H^+ = H_2O$$
$$HCO_3^- + H^+ = H_2CO_3$$
$$CO_3^{--} + 2H^+ = H_2CO_3$$

In addition to P and M values, a B value is of some importance in the operation of boilers. It represents the concentration of free hydroxyl ion in a sample of water, as distinguished from that produced by the hydrolysis of salts of weak acids. The B value is determined by adding neutral barium chloride solution to a measured sample and titrating to the end point of phenolphthalein. P and M values are expressed in terms of equivalent ppm $CaCO_3$, but B values are in terms of parts of CO_3^{--} per 100,000. The B value is converted to ppm $CaCO_3$ by multiplying by 16.6; the result then compares with $(2P - M)$. The use of equivalent ppm $CaCO_3$ as a unit for alkalinity and hardness facilitates the calculation of dosages in a number of operations in water treating and simplifies the comparison of analyses of different waters.

Table 5.4 summarizes the relationships among P and M values and the concentrations of bicarbonate, carbonate, and hydroxyl ions.

In general, free hydroxyl ion is formed in boiler water by the decomposition of carbonates and bicarbonates introduced in the feed water.

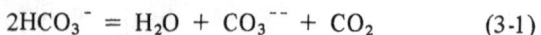

$$2HCO_3^- = H_2O + CO_3^{--} + CO_2 \qquad (3\text{-}1)$$

$$CO_3^{--} + H_2O = CO_2 + 2OH^- \qquad (2\text{-}13)$$

TABLE 5.4

Alkalinity Relationships

Condition	CaCO$_3$ (ppm)		
	(OH$^-$)	(CO$_3^{--}$)	(HCO$_3^-$)
P = 0	0	0	M
P < M/2	0	2P	(M − 2P)
P = M/2	0	2P	0
P > M/2	(2P − M)	2(M − P)	0
P = M	M	0	0

Hydroxyl ion thus formed normally produces a pH in the range of 11-12; this condition is called the free caustic or caustic reserve system of alkalinity control. It has the advantage of providing a considerable amount of neutralizing capacity in the event of an acid leak, but it is essential that internal surfaces be kept clean and free of porous deposits, as these promote caustic gouging, the most common form of corrosion in high-pressure boilers. This phenomenon is discussed in Chapter 2, pp. 21-22.

When feed water is obtained from demineralizers, deaerated condensate, or evaporators it is necessary to add sodium hydroxide to the boilers to produce the necessary alkalinity. In boilers treated by the caustic reserve system it is customary to maintain the total alkalinity (M value) at 10-20 percent of the total dissolved solids. The M-alkalinity can be reduced, for instance, by adding sulfuric acid to the make up water before passing it through cation exchangers. The effect of this is to increase the ratio of total solids to total alkalinity in the boiler water. Another system for controlling alkalinity, the coordinated phosphate-pH method, was devised especially for preventing caustic embrittlement; it is discussed in Section 5.4b.

b. Oxygen Scavengers

Great care is taken to eliminate dissolved oxygen from boiler feed water because of its tendency to produce pitting in boiler drums. Vacuum deaerators that operate cold reduce the concentration of oxygen to 0.4-1.4 ppm; open deaerating heaters reduce it to 0.3-0.7 ppm; spray-type deaerating heaters with two or three stages yield an effluent containing less than 0.01 ppm of dissolved oxygen. The corrosive effect of very small amounts of oxygen on the boiler drum is negligible. Overloaded deaerators often furnish water containing higher concentrations of oxygen, however, so an excess of a reducing agent such as sodium sulfite or hydrazine is maintained, both in the feed water and in the boiler water, to serve as an oxygen scavenger.

Hamer, et al.[15] have shown that as water containing dissolved oxygen enters the steam drum most of it flashes immediately into the steam space. Table 5.5 lists the ratios of oxygen in boiler water to that in feed water at several different pressures.

TABLE 5.5

Distribution of Oxygen in Feed and Boiler Water

Pressure (psi)	ppm O_2 in boiler water/ppm O_2 in feed water
180	1/5000
600	1/1570
1000	1/950
2000	1/500

The ratios in Table 5.5 show that if residual oxygen is to be effectively reduced the reducing agent must be added early in the preboiler system so that reduction takes place before oxygen reaches the steam drum. A residual of oxygen scavenger is always maintained in the boiler to ensure protection during upsets.

Sodium Sulfite. Sulfite reacts with oxygen according to the following equation:

$$\tfrac{1}{2}O_2 \; + \; SO_3^{\;--} \; = \; SO_4^{\;--} \tag{5-9}$$

This reaction is rapid at high temperatures, but it is slow at ordinary temperatures. Sodium sulfite solution, therefore, is often added continuously to the storage section of deaerators, or injected into the feed water line ahead of economizers and preheaters. To make the preboiler protection more effective catalysts are added to the salt, which is then sold as catalyzed sodium sulfite. Pye[16] discusses the effect of small concentrations of copper and cobalt ions upon the rate of reaction in Eq. (5-9). Cobalt is the more efficient catalyst: 10 ppm of dissolved oxygen was reduced to zero in 18 s by an excess of sulfite in the presence of 0.01 ppm of Co^{++}. Commercial sodium sulfite for oxygen scavenging contains about 0.25 percent of cobaltous chloride ($CoCl_2 \cdot 6H_2O$). Copper is never used because of the destructive action of cupric ions on steel boiler tubes. When using chelants discussed in Section 5.2, it is essential that catalyzed sodium sulfite be added separately, for if it is mixed with the chelant the metallic ions are inactivated as oxidation catalysts. Table 5.6 shows suggested ranges of concentrations of sodium sulfite in boilers operated at various pressures.

TABLE 5.6

Suggested Ranges for Sodium Sulfite

Maximum pressure (psi)	ppm Na_2SO_3
200	80–90
300	60–70
450	45–60
600	30–45
750	25–30
900	15–20
1000	NR
1500	NR

NR = not recommended

In modern practice sodium sulfite is not recommended for boilers operated at more than 900 psi, because of the decomposition of sulfite to hydrogen sulfide and sulfur dioxide.

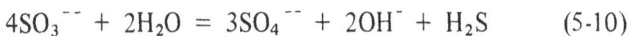

$$4SO_3^{--} + 2H_2O = 3SO_4^{--} + 2OH^- + H_2S \qquad (5\text{-}10)$$

$$SO_3^{--} + H_2O = 2OH^- + SO_2 \qquad (5\text{-}11)$$

Alexander and Rummel[17] have published a comprehensive discussion of sodium sulfite and its decomposition products at various temperatures, pressures, and concentrations. They found that appreciable increases in concentration of sodium sulfite in boiler water at 950-1050 psi lowered the pH and increased the conductivity of condensate. Thus, at 1050 psi the pH of condensate was 8.3 with 3 ppm of Na_2SO_3, but fell to 5.9 with 57 ppm of Na_2SO_3 in the boiler water. They found that the rate of decomposition was negligible at 530-535 F, but it increased rapidly up to 600 F.

Eq. (5-9) shows that 7.88 lb of Na_2SO_3 reacts with 1 lb of oxygen to yield 8.88 lb of Na_2SO_4, thus increasing the total dissolved solids in boiler water. Also, Fiss[18] has objected to the tendency of sodium sulfite solutions to seep around hand hole gaskets, valve stems, and pump glands. Solutions of this salt have a wetting ability that permits them to leak through normally tight connections, particularly screwed joints.

Hydrazine. Hydrazine is the preferred oxygen scavenger in boilers operated above 900 psi. The overall reaction with oxygen is represented by the following equation, which shows that 1 lb of hydrazine reacts with 1 lb of oxygen without increasing the concentration of dissolved solids.

$$N_2H_4 + O_2 = N_2 + 2H_2O \qquad (5-12)$$

Everitt, et al.[19] report that the rate of the reaction in Eq. (5-12) is negligibly slow below 392 F in the absence of a catalyst. Surface catalysts such as copper gauze or an activated carbon filter in the feed line increase the reaction rate of hydrazine with oxygen. More recently, hydroquinone has been introduced as an oxidation catalyst in some formulations.

Stones[20] points out that the reduction of oxygen by hydrazine probably follows a heterogeneous reaction mechanism. It is a common experience to find that two to four weeks elapse after initiating hydrazine treatment before a residual can be detected in the boiler water. This is because reactions such as Eqs. (5-13) and (5-14) are rapid and consume hydrazine.

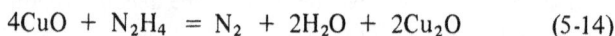

$$6Fe_2O_3 + N_2H_4 = N_2 + 2H_2O + 4Fe_3O_4 \qquad (5-13)$$

$$4CuO + N_2H_4 = N_2 + 2H_2O + 2Cu_2O \qquad (5-14)$$

It is likely that the actual reduction of oxygen takes place in two steps involving the oxidation and reduction of oxides of iron.

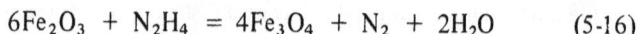

$$4Fe_3O_4 + O_2 = 6Fe_2O_3 \qquad (5-15)$$

$$6Fe_2O_3 + N_2H_4 = 4Fe_3O_4 + N_2 + 2H_2O \qquad (5-16)$$

The oxide debris in boilers treated with hydrazine is invariably black, and discontinuing the addition of hydrazine does not immediately lead to the presence of oxygen in the steam system. Furthermore, water treated with hydrazine rarely contains any significant amount of suspended oxides.

Hydrazine hydrate is available as an 85 percent solution, but it is safer to handle a 35-40 percent solution, which does not flash. The solutions, which are strongly alkaline, are shipped in 18/8 stainless steel returnable drums. Like sodium sulfite, hydrazine is best added to the storage section of the deaerator; treatment is started at 3-5 times theoretical. On ships, hydrazine should be added to the storage section of the deaerator when in port, but at sea it is desirable to add it to the turbine cross-over. In the first instance advantage is taken of long residence time; in the second the heat of the exhaust steam from the first stages of the turbine hastens reaction with oxygen. The turbine cross-over, of course, is upstream of the deaerator, but the loss of hydrazine there is small, i.e., less than 5 percent. Samples for the determination of hydrazine residual had best be taken from the feed water line after the last preheater stage rather than from the boiler blowdown; the residual should be 0.002-0.005 ppm.

Table 5.7 lists the residuals of hydrazine that can be maintained at various operating pressures; concentrations in excess of the upper values in the table decompose to ammonia. Increasing the feed rate of hydrazine, therefore, increases the amount decomposed and the concentration of ammonia in the steam.

TABLE 5.7

Equilibrium Concentrations of Hydrazine

Drum pressure (psi)	Residual hydrazine (ppm)
900	0.10-0.15
1000	0.10-0.15
1500	0.05-0.10
2000	0.05-0.10
2500	0.02-0.03
3000	0.01-0.02

Stones[20] has studied the decomposition of hydrazine to ammonia.

$$2N_2H_4 = N_2 + H_2 + NH_3 \qquad (5\text{-}17)$$

This reaction begins at 400 F and is virtually complete at 625 F. For this reason only very low residuals are found in boiler water and no hydrazine survives passage through the superheater. One or two parts per million of ammonia accumulate in the steam-condensate system assisting in keeping the condensate alkaline. In the event of faulty deaeration, however, ammonia in the presence of oxygen attacks condenser tubes, drain coolers, and feed water preheaters, all of which are likely to be fabricated with copper alloys.

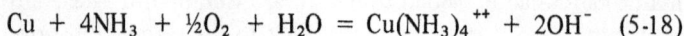

$$Cu + 4NH_3 + \tfrac{1}{2}O_2 + H_2O = Cu(NH_3)_4^{++} + 2OH^- \quad (5\text{-}18)$$

The corrosive effect of oxygen on copper is greatly enhanced in the presence of ammonia because of the stability of the cupric ammonia complex ($K = 4.7 \times 10^{-15}$). Copper ion subsequently enters the boiler, plates on the steel tubes where it becomes a cathode, and the underlying steel corrodes. In this coupled system iron gives cathodic protection to the plated copper, but is itself sacrificed. Many tube failures can be attributed to this sequence of events. This discussion of oxygen scavengers is concluded by summarizing the advantages and disadvantages of each.

Sodium Sulfite

   Advantages:
      1. Catalyzed product reacts rapidly.
      2. Residual is easy to determine.
      3. Not difficult or hazardous to handle.

   Disadvantages:
      1. Increases dissolved solids in boiler water.
      2. Relatively high equivalent weight.
      3. Decomposes to acidic gases at high temperature.

Hydrazine

   Advantages:
      1. Low equivalent weight.
      2. Does not increase dissolved solids.

   Disadvantages:
      1. Vapor toxic.
      2. Contact with solutions causes dermatitis.
      3. Excess decomposes to ammonia.

4. Reacts slowly with oxygen below 350 F.
5. Concentrated base is flammable.

5.4 PREVENTION OF STRESS CORROSION CRACKING

Stress corrosion cracking occurs when the grain boundaries in steel become filled with oxides. This happens when the metal is stressed beyond its elastic limit while in contact with concentrated alkali. Thus, it is possible that improper welding or failure to stress-relieve in a boiler using the caustic reserve system discussed in Section 5.3a above could lead to stress corrosion cracking (caustic embrittlement). Two chemical approaches are used to avoid this type of corrosion: specific inhibitors can be added to control intergranular cracking, or if practicable, the coordinated phosphate-pH method can be applied for controlling alkalinity. Careful attention to stress-relieving, however, is the most effective preventive.

a. Inhibitors

Although the mechanism is unclear, certain chemicals when added to boiler water inhibit caustic embrittlement. Of these the most effective is sodium nitrate, although it suffers from the disadvantage of being difficult to measure and of being subject to decomposition. It decomposes on heating to sodium nitrite, which in alkaline solution is then successively reduced by magnetite or ferrous hydroxide to hyponitrite, $N_2O_2^{--}$, hydroxylamine, NH_2OH, and finally to ammonia, NH_3. Requirements are high on account of this decomposition. Also, the presence of free alkali is necessary, for as Uhlig[21] mentions, stressed steel may fail in hot nitrate solutions when alkali is absent. The recommended dosage is 20-40 percent of the sodium hydroxide alkalinity, i.e., $(2P - M)$. The desired ratio is

$$\frac{NaNO_3}{NaOH} = \frac{ppm\ NO_3^-\ \times\ 1.36}{(2P - M)\ \times\ 0.80} \qquad (5\text{-}19)$$

which should fall in the range of 0.20-0.40.

Certain condensed tannins, e.g., quebracho, seem to have merit as inhibitors of caustic embrittlement, although Purcell and Whirl[22] note that some brands are good inhibitors while others are worthless. Commercial formulations are nearly always mixtures that may include catechol types such as quebracho and cutch tannins, as well as the pyrogallol class, which includes chestnut. All of these are easy to apply and measure; the usual dosage is 150–200 ppm.

For many years various sulfate-alkalinity ratios were recommended for preventing caustic embrittlement. These were based on a study by Parr and Straub[23] in which they observed that certain local waters in Illinois that were high in bicarbonate and low in sulfate caused stress corrosion cracking. Among other things they measured the effect of increasing the $Na_2SO_4/NaOH$ and $Na_2CO_3/NaOH$ ratios. This led to differing opinions of which ratio should be used and what its value should be. One of these, for example, was that $Na_2SO_4/NaOH$ should exceed 2.5.

$$\frac{Na_2SO_4}{NaOH} = \frac{ppm\ SO_4 \times 1.42}{(2P - M) \times 0.80} \tag{5-20}$$

Another stipulated that Na_2SO_4/Na_2CO_3 should > 3.

$$\frac{Na_2SO_4}{Na_2CO_3} = \frac{ppm\ SO_4^{--} \times 1.42}{2(M - P) \times 1.06} \tag{5-21}$$

Still a third specification (for 850-psi boilers) was that $0.083 \times$ ppm SO_4^{--}/B should be in the range of 3.8–4.5. This last ratio has the unusual units of ppm SO_3/ppm CO_3^{--}.

$$\frac{0.083 \times ppm\ SO_4}{10 \times B} = \frac{ppm\ SO_3}{ppm\ CO_3^{--}} \tag{5-22}$$

One theory held that sodium hydroxide reduced the solubility of sodium sulfate, causing it to precipitate in leaking seams, and thus prevent further concentration of alkali. In any event many subsequent studies have confirmed that the sulfate/alkalinity ratio, however calculated, has no effect in preventing stress corrosion cracking. In fact, maintaining these ratios at the recommended values in boilers over 900 psi often causes turbulence and carryover.

b. Coordinated Phosphate-pH Control of Alkalinity

As previously mentioned, the presence of concentrated free hydroxyl ion is necessary before caustic embrittlement can occur. Purcell and

Whirl,[22] noting that solutions of trisodium phosphate are alkaline enough by hydrolysis to prevent corrosion, but do not leave a residue of free sodium hydroxide upon evaporation, proposed a method for controlling alkalinity by coordinating phosphate concentration and pH.

When trisodium phosphate dissolves in water the phosphate ion, on account of its strong electrical field, is about 95 percent hydrolyzed to HPO_4^{--}, as in Eq. (5-23). When the solution is evaporated, however, Eq. (5-23) is reversed, leaving only Na_3PO_4 as a residue.

$$PO_4^{---} + H_2O = HPO_4^{--} + OH^- \qquad (5\text{-}23)$$

The hydrolysis constant, K_h, for this reaction can be calculated from the third dissociation constant of phosphoric acid and the ion product of water.

Blackburn[24] quotes $pK_3 = 12.360$ for phosphoric acid; this value is incorporated here. Thus,

$$K_h = K_w/K_3 = 10^{-14}/4.36 \times 10^{-13} \qquad (5\text{-}24)$$

$$= 2.29 \times 10^{-2}$$

From this value the pH of any known solution of trisodium phosphate can be calculated and thus a curve relating pH and total phosphate can be plotted. Suppose there is a solution of trisodium phosphate containing the equivalent of 100 ppm of total phosphate. The corresponding pH is calculated as follows:

$$100 \text{ ppm } PO_4^{---} = (100 \times 10^{-3}/95) \text{ mol/l}$$

$$= 1.05 \times 10^{-3} \text{ mol/l}$$

Using the expression for the hydrolysis

$$\frac{(HPO_4^{--})(OH^-)}{(PO_4^{---})} = 2.29 \times 10^{-2} \qquad (5\text{-}25)$$

and noting that when trisodium phosphate dissolves in water equal concentrations of (HPO_4^{--}) and (OH^-) are produced, i.e.,

$$(HPO_4^{--}) = (OH^-)$$

then

$$\frac{(OH^-)^2}{(PO_4^{---}) - (OH^-)} = 2.29 \times 10^{-2} \qquad (5\text{-}26)$$

Substituting the value of total phosphate and rearranging gives

$$(OH^-)^2 + 2.29 \times 10^{-2}(OH^-) - 1.05 \times 2.29 \times 10^{-5} = 0$$

Solving by the quadratic formula

$$(OH^-) = \frac{-2.29 \times 10^{-2} + \sqrt{(2.29 \times 10^{-2})^2 + 9.6 \times 10^{-5}}}{2} \qquad (5\text{-}27)$$

$$= \frac{-2.29 \times 10^{-2} + \sqrt{6.20 \times 10^{-4}}}{2}$$

$$= (2.49 - 2.29) \times 10^{-2}/2$$

$$= 1.00 \times 10^{-3}$$

$$(H^+) = 10^{-14}/10^{-3}$$

$$= 10^{-11}$$

Whence

$$pH = 11.0$$

Carrying through the arithmetic for a solution containing 10 ppm PO_4^{---} $(1.05 \times 10^{-4}$ mol/l) it is found that the corresponding pH is 9.4. If now ppm PO_4^{---} is plotted on the logarithmic scale against pH on the linear scale on a sheet of semilogarithmic graph paper a straight line as shown in Fig. 5.1 is obtained. In any boiler containing phosphate and caustic the P- and M-alkalinities can be interpreted as follows: If $2P > M$, the two coordinates, total phosphate, and pH define a point to the right of the line, indicating that excess hydroxyl ion is present. If $2P = M$, the two coordinates fall on the line. When $2P < M$ the coordinates define a point to the left of the line, indicating excess disodium phosphate. Also, the equivalent concentration of excess disodium phosphate can be calculated

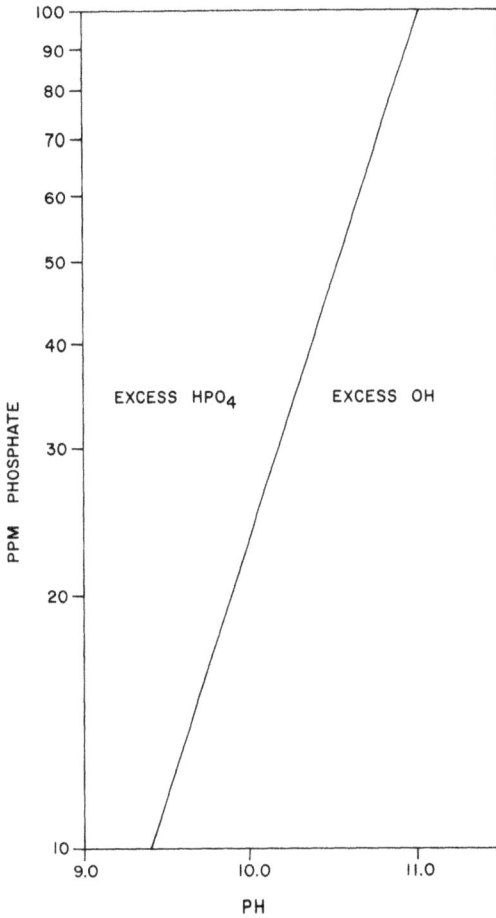

Fig. 5.1

from the expression

$$\text{ppm } Na_2HPO_4 = (M - 2P) \times 0.02 \times 142 = (M - 2P) \times 2.84$$

When operating a boiler on the coordinated phosphate-pH program, conditions are adjusted so that the two coordinates fall just to the left of the line. Thus, there is always some capacity to absorb caustic. This system is particularly suited for marine boilers, which use evaporated water for make up, and for stationary boilers that use demineralized water for make up. The system is not practicable, however, with feed waters containing a great deal of carbonate alkalinity because the latter is converted to free alkali in the boiler requiring excessive amounts of disodium phosphate. Another form of the coordinated phosphate-pH curve is exhibited in Fig. 6.1.

In most feed water the P-alkalinity is less than one-half the M-alkalinity, in which case $CO_3^{--} = 2P$, and $HCO_3^- = (M - 2P)$. The following reactions occur in the boiler:

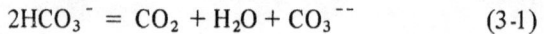

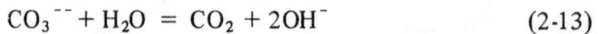

$$2HCO_3^- = CO_2 + H_2O + CO_3^{--} \qquad (3\text{-}1)$$

$$CO_3^{--} + H_2O = CO_2 + 2OH^- \qquad (2\text{-}13)$$

Total carbonate is the sum of that initially present and that formed from bicarbonate.

Initially present: ppm CO_3^{--} = 2P × 0.60

From bicarbonate: ppm CO_3^{--} = (M − 2P) × 1.22 × 60/2 × 61

$$\text{Total carbonate} \quad = \quad \left[\frac{(M - 2P) \times 1.22 \times 60}{2 \times 61} + 2P \times 0.60 \right]$$

Of this total carbonate, approximately 80 percent is hydrolyzed to hydroxyl ion according to Eq. (2-13). The ppm hydroxyl ion produced is

$$\left[\frac{(M - 2P) \times 1.22 \times 60}{2 \times 61} + 2P \times 0.60 \right] \times 0.80 \times 34/60 = \text{ppm } OH^-$$

This expression reduces to

$$M \times 0.272 = \text{ppm } OH^-$$

Also,

$$M \times 0.272 \times 2.94 = M \times 0.80 \text{ ppm } CaCO_3 \text{ equivalent to } OH^-$$

When using the coordinated phosphate-pH method for alkalinity control sufficient disodium phosphate must be added to neutralize the hydroxyl ion formed according to Eq. (2-13).

$$HPO_4^{--} + OH^- = PO_4^{---} + H_2O \qquad (5\text{-}28)$$

As the equivalent weight of disodium phosphate in Eq. (5-28) is 142, the concentration of disodium phosphate required in the feed water to neutralize the hydroxyl ion formed in the boiler water is calculated as follows:

$$\text{ppm } Na_2HPO_4 \;=\; \frac{M \times 0.272 \times 142}{17}$$

$$= M \times 2.27$$

If, for example, a boiler were being fed with water having an M-alkalinity of 3 ppm $CaCO_3$, and operating at 5 percent blowdown, the ppm Na_2HPO_4 required in the feed water is 3×2.27, or 6.8 ppm. This would produce a phosphate concentration in the boiler of

$$\frac{6.8 \times 95 \times 100}{142 \times 5} \;=\; 91 \text{ ppm}$$

which is too high a concentration to keep in solution in most boilers. In practice, of course, phosphate is usually injected into the steam drum, but the total amount required is the same. Thus, if the feed rate is 500,000 lb/h, disodium phosphate must be added at the rate of 3.4 lb/h, either to the feed water, or to the steam drum. If phosphate is to be added to the feed water it is advantageous to use hexametaphosphate, as it does not precipitate as calcium phosphate in feed lines, and it also has a lower equivalent weight than disodium phosphate. The requirement of $(NaPO_3)_6$ is

$$\text{ppm } (NaPO_3)_6 \;=\; M \times 1.63$$

In addition to the coordinated phosphate-pH method, there are two variations called precision control[25] and congruent control.[26] In the former, phosphate is kept in the range of 2-4 ppm with hydroxyl ion, $(2P - M)$, at 15-50 ppm. The purpose of these limits is to prevent the precipitation of $Mg_3(PO_4)_2$, which has a pronounced tendency to stick to hot tubes. At low concentrations of phosphate the less adherent $MgSiO_3$ is preferentially precipitated; also, there is no tendency for Na_3PO_4 to sludge. Ravich and Shcherbakova[27] have shown that crystals that separate from solutions of Na_3PO_4 at high temperatures contain some Na_2HPO_4 and the supernatant liquid contains some NaOH. Solutions in which the composition of the crystals was identical with that of the solution they called congruent compositions. At 572 F congruent composition is obtained when the ratio of sodium to phosphate is 2.85; at 689 F the ratio is 2.65. Marcy and Halstead[26] state that a ratio of $Na/PO_4 = 2.6$ is safe at any pressure. The proper ratio is obtained by mixing H_3PO_4, NaH_2PO_4, Na_2HPO_4, Na_3PO_4, and $NaOH$. For example, 0.6 mole of Na_3PO_4 combined with 0.4 mole of Na_2HPO_4 gives a mixture in which $Na/PO_4 = 2.6$. In the congruent treatment total phosphate is kept in the range of 4-6 ppm with pH at 8.5-9.3.

A summary of the advantages and disadvantages of the coordinated phosphate-pH methods of alkalinity control as compared to the caustic reserve system is provided below.

Advantages:
1. Prevents stress corrosion cracking.
2. Produces a lower concentration of dissolved solids.
3. No risk of caustic gouging under scale or oxide deposits.
4. Absence of caustic minimizes carryover of salts.
5. Turbine deposits are less adherent.

Disadvantages:
1. Close chemical control is required.
2. Little reserve alkalinity in case of mishap.
3. In case of hardness leakage magnesium phosphate precipitates rather than magnesium hydroxide causing scaling.
4. Silica concentration must be kept low (3-4 ppm) as there is no free alkali to prevent silicate scaling and steam distillation of silicic acid.
5. The M-alkalinity of the feed water must be very low (1-2 ppm).

5.5   PREVENTION OF SILICA CARRYOVER AND SCALING

The tendency of silicates to form scales on boiler tubes requires that its concentration in feed water for boilers be reduced as much as possible. At pressures above 500 psi silica steam-distills,[28] forming scales of varying degrees of hardness in superheater tubes and on turbine blades. The efficiency of a turbine is drastically reduced, even by a relatively thin scale. Coulter, et al.[29] have studied the phenomenon of silica carryover using an electrically heated test boiler capable of operating at pressures up to 3200 psi at a maximum steaming rate of 100 lb/h and a heat flux of 20,000 Btu/h/ft^2. An air-cooled condenser and sampling system permitted

TABLE 5.8

Volatility of Silica as a Function of Pressure

Pressure (psi)	SiO_2 in steam (ppm)
500	0.02
1000	0.06
1500	0.30
2000	1.40
2500	6.00
3000	25.00
3200+	50.00

simultaneous collection of water and condensed steam. To measure the extent of mechanical carryover, radioactive phosphorus, P^{32}, in the form of orthophosphate was added to the boiler water. They then measured the level of radioactivity in the condensed steam. This extremely sensitive method showed that some mechanical carryover always occurs, even at very low steaming rates. The amount of silica introduced into steam by mechanical carryover, however, is negligible at pressures above 500 psi compared to that volatilized, for the latter increases logarithmically with increasing pressure.

a. Volatility of Silica

Internal treatment for controlling residual silica introduced in the feed water is, in the main, directed toward preventing it from volatilizing in the steam. If the concentration of silica in steam exceeds about 0.03 ppm troublesome deposits are sure to form in turbines. The volatility of silica is proportional to its concentration in the boiler water and the operating pressure, and inversely proportional to the pH of the water. Table 5.8 shows the concentraion of silica found in steam at several pressures, up to the critical pressure, when the boiler water contains 50 ppm of SiO_2.

Because turbine scales are so troublesome many operators prefer to be conservative with respect to silica concentrations. Table 5.9 lists operating pressures and the corresponding concentrations of silica in the boiler water that give a concentration in the steam approximately 0.02 ppm SiO_2, a level that is ordinarily tolerable.

TABLE 5.9

Maximum Silica in Boiler Water

Pressure (psi)	ppm SiO$_2$ (maximum)
100	250.00
200	200.00
300	175.00
500	40.00
600	35.00
750	30.00
900	20.00
1000	10.00
1500	3.00
2000	1.00
2500	0.50
3200+	0.02

The form in which silicates volatilize is uncertain. Straub and Grabowski[30] consider that meta-monosilicic acid, H_2SiO_3, is the volatile compound, while others favor SiO_2 on account of the slight conductivity of condensates containing as much as several hundred parts per million SiO_2. The first dissociation constant of meta-monosilicic acid, however, is very small ($\approx 10^{-10}$) from which it can easily be calculated that a 300-ppm solution of it is only 3.2×10^{-3} percent dissociated, and consequently would have negligible conductivity. Also, deposits of sodium disilicate are found in the hottest stages of turbines, further complicating the picture.

Splittgerber[31] observed that the addition of various salts, including sodium chloride, sodium sulfate, and trisodium phosphate to boiler water causes the concentration of silica in the steam to decrease. He assumed that silica leaves the water as meta-monosilicic acid according to Eq. (5-29).

$$SiO_3^{--} + 2H_2O = H_2SiO_3 + 2OH^- \qquad (5\text{-}29)$$

Conversely, it would be expected that high alkalinity would reverse this reaction. Table 5.10, from the data of Straub and Grabowski,[30] shows the effect of alkalinity on the volatility of silicic acid in a boiler operated at 1545 psi.

TABLE 5.10

Effect of pH on the Volatility of Silicic Acid

pH	SiO$_2$ in steam/SiO$_2$ in water (%)
12	0.55
11	0.77
10	1.02
9	1.25
8	1.48
7	1.73

Because of the great effect of alkalinity on the volatility of silicic acid, only small concentrations of silica can be tolerated in boilers under the coordinated phosphate-pH method of alkalinity control, e.g., 3 ppm SiO$_2$, maximum at 900 psi. The values in Table 5.9 assume the presence of free caustic with B-readings in the range of 6-12 at pressures < 900 psi.

b. Chemistry of Silica Scaling

Latimer[32] states that in aqueous solution silicic acids participate in complex equilibria on account of the small differences in free energy between the numerous polyacids and their salts. Thus, a variety of mineral types have been identified in boiler scales and deposits. Acmite, $Na_2O \cdot Fe_2O_3 \cdot 4SiO_2$, and analcite, $Na_2O \cdot Al_2O_3 \cdot 4SiO_2$, both of which are extremely insoluble, require high temperature to form and, thus, are seldom found in boilers operated at less than 300 psi. On the other hand, serpentine, $3MgO \cdot SiO_2 \cdot 2H_2O$, and hydrated magnesium orthodisilicate, $Mg_3Si_2O_7 \cdot 2H_2O$, can form at any pressure. Garrels and Christ[33] show that magnetite is unstable relative to ferrous metasilicate, $FeSiO_3$, so that if an aqueous solution contains sufficient silica to satisfy all of the iron present, ferrous metasilicate forms in preference to magnetite. Iron silicate scales are exceedingly insoluble and adherent, but they can be avoided by maintaining adequate caustic alkalinity, and by limiting the concentration of iron in the feed water.

In studying the analytical methods for determining silica in feed water, Dwyer and Frith[34] found that neither colloidal silica nor iron are detected by the usual colorimetric chemical methods. Thus, more silica and iron may be introduced into a boiler than the ordinary control procedures indicate. This is particularly significant in operating supercritical boilers in which it is necessary to limit the concentration of iron and silica to 10-20 ppb in the boiler water.

In low-pressure boilers ($\leqslant$ 200 psi) there is no tendency for silica to steam-distill. It does form sludge and scale, however, and Gray[35] has proposed an empirical formula for controlling these using an internal treatment with sodium carbonate. The success of the treatment depends upon two factors: the total carbonate alkalinity in the boiler must be around 300 ppm $CaCO_3$; and the concentrations of calcium hardness, magnesium hardness, and silica in the feed water must be such that

$$(100/CaH)[(MgH/3) - SiO_2] > 7 \qquad (5\text{-}30)$$

These conditions ensure that the total hardness in the boiler will be $<$ 5 ppm, and that magnesium silicate will precipitate as a nonadherent sludge. It is readily apparent that in most waters the proper relationship can be obtained by adding a magnesium salt to the feed water.

5.6   CLARIFICATION OF BOILER WATER

Unless the amount of material precipitated in a boiler by internal treatment is small, it is advisable to add dispersants to prevent insoluble crystals from growing into aggregates. If their growth is unimpeded, sludge accumulates in areas where the flow of water is slow and interferes with circulation. At high pressures sludge may bake on boiler tubes or form porous deposits that can cause corrosion within the boiler. For many years naturally occurring organic compounds that form colloidal aqueous solutions have been used to condition or "fluidize" insolubles in boiler water. Recently more efficient synthetic polymers have been developed that disperse insoluble precipitates. In general, the application of polymers is limited to pressures below 1000 psi. Feed water required at higher pressures necessarily contains negligible concentrations of suspended solids, dissolved iron, and hardness. Also, some of the organic polymers are unstable at high temperature.

a.   Natural Organic Dispersants

Condensed tannins, which are polymeric glycosides of gallic acid, are useful coagulants in boilers at low pressures containing 300-400 ppm of suspended solids. Tannin forms a dense, nonadherent floc with calcium carbonate and magnesium hydroxide that can be eliminated from the boiler by intermittent blowdown. When colloidal particles are present, tannin encloses them in a protective envelope that prevents their coagulation by electrolytes.

Sulfonated lignins, prepared by treating wood pulp with sodium bisulfite, are economical dispersants for phosphates and iron oxides. They too are protective colloids that act by coating particles to form a clarified colloidal solution. Upon treating sulfite pulp with chloroacetic acid and alkali, sodium carboxymethylcellulose is obtained, which also forms colloidal solutions in water. High concentrations are required to produce any significant dispersion and consequently this treatment is expensive. Other natural polymers that are used as sludge conditioners include starch, quebracho, pyrogallol, sodium alginate, and sodium mannuronate. Starch is particularly effective for coagulating silica and oil at low pressures.

Denman and Salutsky[36] found that sulfonated lignins tend to form carbonized deposits in sections at high temperature; these products begin to char at about 500 F. Holmes and Jacklin[37] report that starch, sulfonated lignin, chestnut tannin, and quebracho applied singly reduced the volume of deposits on the evaporative surface in an experimental boiler at 800 psi. Although the organic molecules showed no tendency to decompose, these authors recommend caution in using them.

In general, all of these organic materials are most effective when used in combinations rather than singly. Various recommendations for their application have been published, but the following dosages are typical: at 200 psi, or less, 150 ppm; at 500 psi, 70 ppm; at 800 psi, 3-10 ppm. The conditioners can be conveniently added with phosphate.

b. Synthetic Polymers

The function of a dispersant is to promote the suspension of insoluble salts, especially calcium and magnesium phosphates, in the bulk of the circulating water rather than allowing them to precipitate on surfaces where heat is transferred. After an insoluble particle has formed the polymer enters into a surface reaction with it that prevents or interferes with further growth of the crystal. Two anionic polymers (polyelectrolytes) that have been used to disperse insolubles in boilers are polyacrylate and polymethacrylate.

Polyacrylate with molecular weight in the range of 5000-10,000 is sometimes described as an antiprecipitant because a complex (not a chelonate) forms between calcium ion and the carboxyl groups.[38] In the presence of polyacrylate deformed crystals (somatoids) of calcium sulfate are obtained similar to those described by Buehrer and Reitemeier[39] in their studies of the effect of polyphosphate on the crystallization of calcium carbonate.

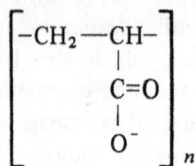

$$
\begin{bmatrix} -CH_2-CH- \\ \quad\quad | \\ \quad\quad C{=}O \\ \quad\quad | \\ \quad\quad O^- \end{bmatrix}_n
\qquad
\begin{bmatrix} \quad\quad CH_3 \\ \quad\quad | \\ -CH_2-C- \\ \quad\quad | \\ \quad\quad C{=}O \\ \quad\quad | \\ \quad\quad O^- \end{bmatrix}_n
$$

Polyacrylate   Polymethacrylate

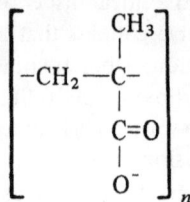

Zeleny and Vithani[40] used an experimental boiler to evaluate the effects of methacrylic acid, polymethacrylic acid with molecular weight in the range of 6500-12,400, and pyrogallol on the rate of scaling by calcium phosphate of a heating surface at 500 psi. Some of their results are tabulated in Table 5.11. These values show that scaling rate increases with heat flux with or without conditioner; that polymethacrylic acid is superior to the monomer; and that pyrogallol is the least effective of the three additives. Pyrogallol at this pressure gives poorer results than no additive, probably because of the charring at high temperature mentioned in Section 5.6a.

Zeleny and Vithani[40] suggest that the presence of hydroxyl groups in organic additives plays a role in preventing the formation of scale by providing more favorable nucleation sites for the formation of crystals of calcium phosphate in the boiling water than are available on the heating surface. Denman and Salutsky[36] found that in the pressure range from 90 to 600 psi, sodium polyacrylate outperforms sodium polymethacrylate except at low molecular weights, and that both are superior to sulfonated lignin, chestnut tannin, and sodium alginate. Neither of the synthetic polymers show any sign of decomposition up to 600 F. At 700 F charring is evident; at 1000 F charring is complete. Polyacrylate at 6 ppm prevents the accumulation of "oil balls" and the formation of sticky oil sludges and oily heating surfaces. Satisfactory dosages of polyacrylates, in general, are 10-15 ppm below 500 psi, and 5-10 ppm from 500 to 900 psi.

TABLE 5.11

Effect of Organic Additives on Scaling

Additive	Concentration (ppm)	Heat flux (Btu/ft^2-h)	Scaling rate (g/cm^2-h)
None	0	53,000	21.6
None	0	58,000	28.2
None	0	79,000	31.1
Methacrylic Acid	70	49,000	7.0
Methacrylic Acid	70	52,000	18.1
Methacrylic Acid	70	88,000	20.4
Pyrogallol	70	–	32.5
Pyrogallol	140	–	140.7
Polymethacrylic Acid	70	–	Negligible

5.7 PREVENTION OF FOAMING IN BOILERS

Foaming in boilers is caused by high concentrations of alkali and salts; by finely divided suspended solids—in particular, oxides of iron and copper; and by colloids such as silt or saponifiable oil. Also, the evolution of bubbles of carbon dioxide released by the decomposition of carbonates produces a layer of foam on the surface of the water in the steam drum similar to that on a glass of beer. The persistence of foam becomes greater as the concentration of salts in the boiler water and the surface tension increase.

Foam is composed of bubbles of steam separated from each other by thin films of boiler water. These are stabilized by repelling forces, principally electrical charges, that arise at the steam-liquid interface. Thinning of the liquid film is opposed by the electrical repulsion of similar charges on both surfaces, which prevents rupturing and coalescence. Hydrophilic colloids such as magnesium hydroxide, micelles of silica, and organic sols can all induce foaming upon being adsorbed at these interfaces. When stable bubbles pile up on the surface of the water in the steam drum, solids dissolved or adsorbed in the water films pass into the steam system and form scales in the superheater and fouling in turbines.

A phenomenon known as "light water" often occurs with foaming, but usually without a stable layer of foam on the surface of the water. This condition arises from an increase in the apparent volume of boiler water caused by the presence of a multitude of small noncoalescing bubbles of steam in the water. The effect is the same as surface foaming in as much as the water has turned into bubbles of steam surrounded by a film of water, which flows into the steam outlet as wet steam contaminated with boiler salines.

a. Mechanism of Foaming

Jacoby and Bischmann[41] have made a photographic study of the formation of steam bubbles on various heating surfaces. In experiments using distilled water at atmospheric pressure, they found that bubbles form on a smooth surface at a few isolated points, then rapidly coalesce as they rise through the liquid. When formed on a surface thinly scaled with calcium carbonate boiling is smoother, more evenly distributed, and the bubbles are smaller. On a heavily scaled surface bubbles form so close together that there is considerable coalescence before they are released from the heating surface. In the extreme case this becomes film boiling, in which a film of vapor separates the heating surface from the liquid and reduces heat transfer. Oily surfaces induce film boiling and overheating of boiler tubes. A saline solution boiling on a scaled surface forms innumerable small bubbles, closely packed as they rise, but with little tendency to coalesce. The volume of the bubbles decreases with increasing concentration of dissolved solids and with decreasing surface tension. Also, as the radius of a steam bubble becomes smaller the rate of its ascent becomes smaller. The surface tension of water decreases as its temperature increases, becoming zero at the critical temperature, 705 F. Inorganic salts increase the surface tension of water as shown in Table 5.12, but almost all organic compounds that dissolve in water lower its surface tension. Sucrose is an exception.

b. Antifoaming Additives

Originally, castor oil emulsion was used in an attempt to control foaming in boilers. It had the serious disadvantage, however, of being rapidly saponified to the sodium soap of ricinoleic acid, which increased the severity of foaming. Gunderson and Denman[42] have described the preparation and use of polyalkylene polyamides, which are essentially nonsaponifiable, as antifoams. These and other foam inhibitors cause steam bubbles to coalesce as they form on heating surfaces, resulting in the release of rela-

TABLE 5.12

Surface Tension of Aqueous Solutions

Solution	Temperature (C)	Solute (ppm)	Surface tension (dynes/cm)
H_2O	20	–	72.75
H_2O	100	–	58.85
NaCl	20	5,850	72.92
Na_2CO_3	20	26,500	73.40
NaOH	18	28,000	74.40
NaOH	18	60,000	75.90
NaOH	18	200,000	83.10

tively large bubbles. The polyamides are prepared by acylating a polyamine with a fatty acid; the simplest of these compounds is distearoylethylenediamide.

$$C_{17}H_{35}CONH-CH_2CH_2-NHCOC_{17}H_{35}$$

An important feature of this structure is the spacing of two polar radicals by two (or more) carbon atoms to establish an effective balance between the hydrophilic and hydrophobic portions of the molecule. Other examples are dioleylpiperazine, dipalmitylpiperazine, N,N'-dihexadecyladipamide, and N,N'-dioctadecylsebacamide.

These compounds owe their activity as antifoams to their capacity for forming hydrogen bonds easily. They are strongly adsorbed at each steam-water interface created by the formation of a bubble where they disrupt the repellent forces that stabilize these bubbles. The polyamides are especially effective in low-pressure boilers that carry high concentrations of dissolved solids. Dosages of 1 ppm collapse surface foam, promote coalescence of ascending bubbles, and prevent carryover, even when dissolved solids are as high as 20,000 ppm. These compounds also function well in the presence of oils or soaps.

When a rising bubble of steam nears the surface of water not prone to foam, a mound of liquid is raised above the bubble, which immediately bursts, emitting a fine spray. This may become entrained in the steam as an aerosol. The polyamides prevent this entrainment when present in the concentration range of 0.2–2.0 ppm. The presence of a protective colloid such as tannin seems to enhance the inhibition of foaming by the poly-amides, so the latter are often prepared in an emulsion with tannins, then dispersed in phosphate or any other alkaline boiler compound that is fed directly to the boiler. The polyamides are also sold as powders, briquettes, or caustic pastes; they are insoluble in water.

Fordyce, et al.[43] have described the synthesis of higher polyoxy-alkylene glycols, examples of which are octadecaoxyethylene glycol, $HO(CH_2-CH_2O)_{16}H$, and dotetracontaoxyethylene glycol, $HO(CH_2-CH_2O)_{42}H$. Compounds of this type are used as antifoams in high-pressure boilers with total dissolved solids < 2000 ppm; they are soluble in cold water, but insoluble in hot water. The usual recommended dosage of the mildly alkaline concentrated commercial solutions is 1-6 ppm; 0.5-1 ppm of the active ingredient is sufficient.

Methylpolysiloxane fluids (silicones) are used primarily in nonaqueous systems, but silicone emulsions are occasionally applied to water, although their cost is high. Other defoamers that have been proposed include hepta-decylthioamide and disulfonamides of long-chain aliphatic sulfonic acids.

5.8 CORROSION INHIBITORS FOR STEAM CONDENSATE SYSTEMS

The principal corrosive agent in condensed steam is carbon dioxide, which is generated in boilers by the thermal decomposition of carbonates and bicarbonates introduced in feed water.

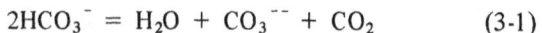

$$2HCO_3^- = H_2O + CO_3^{--} + CO_2 \qquad (3\text{-}1)$$

$$CO_3^{--} + H_2O = CO_2 + 2OH^- \qquad (2\text{-}13)$$

The reaction in Eq. (3-1) is complete and that in Eq. (2-13) is 80 percent complete within six hours at water temperatures above 185 C.[44] This accounts for the observation that in many boilers the P-alkalinity is 80-90 percent of the M-alkalinity.

Carbon dioxide leaves the boiler with the steam and is distributed throughout the steam-condensate system. When steam condenses part of the carbon dioxide dissolves in the condensate where it is hydrated to some extent, forming the weak acid, H_2CO_3, carbonic acid. The latter partially dissociates, releasing a small concentration of hydrogen ion.

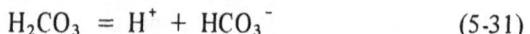

$$H_2CO_3 = H^+ + HCO_3^- \qquad (5\text{-}31)$$

In Table 5.13 are tabulated the pH values calculated from the first dissociation constant of carbonic acid, $K_1 = 4.3 \times 10^{-7}$, corresponding to various concentrations of carbon dioxide in pure water. In these calculations,
the practical assumption that all dissolved carbon dioxide is hydrated to
carbonic acid has been made.

TABLE 5.13

Effect of Carbon Dioxide on pH

ppm CO_2	pH
1.0	5.53
3.0	5.28
5.0	5.17

In actual samples of condensate, however, it is common to find pH
values above eight, even with corrosion rates of several hundred mils per
year. The reason for this is that hydrogen ions produced according to Eq.
(5-31) rapidly oxidize iron, forming hydrogen gas and ferrous ion and leaving a preponderance of bicarbonate ion in the solution.

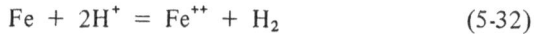

$$Fe + 2H^+ = Fe^{++} + H_2 \qquad (5\text{-}32)$$

If the equilibrium equation for the first dissociation of carbonic acid,
written to introduce the term $1/(H^+)$ is examined, the logarithm of which
is pH, it is seen that the pH of a solution depends upon the ratio of bicarbonate (M-alkalinity) to (H_2CO_3) rather than upon the concentration of
H_2CO_3 alone.

$$\frac{(HCO_3^-)}{(H_2CO_3)} = \frac{(4.3 \times 10^{-7})}{(H^+)}$$

Thus, the reduction of hydrogen ions by iron represented by Eq. (5-32)
continuously increases the ratio of bicarbonate to carbonic acid and raises
the pH. The net effect is perhaps more obvious if Eqs. (5-31) and (5-32)
are combined:

$$Fe + 2H_2CO_3 = Fe^{++} + 2HCO_3^- + H_2 \qquad (5\text{-}33)$$

Carbonic acid, in common with other acids, produces typical gouged or grooved corrosion patterns at particular sites including threaded pipe fittings (elbows, valves, tees), the downstream side of steam traps and control valves where abrupt pressure changes occur, and in locations where non-condensable gases concentrate. Thus, in vertical reboilers with steam on the shell side an obvious thinning, or "necking down," of tubes is often visible at the condensate level. Indeed, the most severe corrosion always occurs at the liquid level, at horizontal baffles where water collects, and on the wet upper surface of bottom tube sheets. Long horizontal runs of return line frequently show characteristic thinning and grooving in the bottom surfaces of the pipes.

Oxygen enters condensate systems through air leaks at threaded joints, through faulty steam traps and packing glands, or it can be forced in through small openings in condensers where the internal pressure is less than that of the atmosphere. Its corrosive action is much more rapid than that of carbon dioxide; in combination they are more virulent than either alone, especially if the pH < 6.

Schenk and Weber[45] have studied the rate of oxidation of ferrous ion by oxygen in a bicarbonate buffered system and found that through the pH range of 6.6–7.1 the rate of oxidation is first order with respect to the concentration of ferrous ion, and second order with respect to the concentration of hydroxyl ion at constant partial pressure of oxygen, i.e.,

$$-d(Fe^{++})/dt = k(Fe^{++})P_{O_2}(OH^-)^2 \qquad (5\text{-}34)$$

The corrosion (rusting) of iron by oxygen alone can be represented as

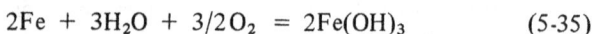

$$2Fe + 3H_2O + 3/2O_2 = 2Fe(OH)_3 \qquad (5\text{-}35)$$

The oxidation of ferrous ion produced by carbonic acid according to Eq. (5-33) proceeds as follows:

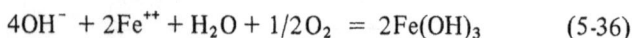

$$4OH^- + 2Fe^{++} + H_2O + 1/2O_2 = 2Fe(OH)_3 \qquad (5\text{-}36)$$

The reactions in Eqs. (5-35) and (5-36) both have a great tendency to go to completion because of the extreme insolubility of $Fe(OH)_3$. From the solubility product, $K_{sp} = 6 \times 10^{-38}$, it can be calculated that the maximum concentration of ferric ion that can remain in solution at pH 6.0 is 3.4×10^{-9} ppm.

Because of the tendency of the gas to come out of solution and form bubbles on the surface of metal, oxygen attack often takes the form of irregular pitting and tuberculation, with an accumulation of corrosion products around the site of attack.

Occasionally small amounts of hydrogen sulfide or sulfur dioxide appear in condensates. At temperatures above 1000 F sulfite has been shown to undergo the following decompositions:

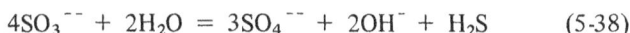

$$SO_3^{--} + H_2O = 2OH^- + SO_2 \qquad (5\text{-}37)$$

$$4SO_3^{--} + 2H_2O = 3SO_4^{--} + 2OH^- + H_2S \qquad (5\text{-}38)$$

These reactions can occur, for example, if boiler water containing sodium sulfite is thrown into the superheater. Hydrogen sulfide is so reactive that most of it remains as a black layer of ferrous sulfide on the superheater tubes. The effect of sulfur dioxide is similar to that of carbon dioxide; that is, it forms a weak acid, H_2SO_3, that lowers the pH of condensate.

Corrosion inhibitors, including ammonia and volatile amines, are available that neutralize carbonic acid and elevate the pH of condensate. Ammonia has some serious disadvantages, however, and when the concentration of carbon dioxide in steam is high the cost of neutralizing amines is prohibitive. Filming amines are much more economical as the dosage required is low and independent of the concentration of carbon dioxide. Ideally, they form a chemisorbed monomolecular layer on metal surfaces that serves as a physical barrier against carbon dioxide, oxygen, and other dissolved gases. These inhibitors have been discussed in a general way by Maguire;[46] they are considered further in the following sections.

a. Ammonia as a Neutralizing Agent

Ammonia was no doubt the first volatile base to be used to minimize corrosion by carbonic acid in condensates. It has the advantages of being inexpensive, having a low equivalent weight, and being relatively easy to handle. The reaction of carbon dioxide with ammonia results in the formation of a coordination covalent bond. If the reaction occurs in the absence of water it can be formulated according to the electronic theory of acids and bases[47] as follows:

$$O_2C + :NH_3 = O_2C:NH_3 \qquad (5\text{-}39)$$

The product of this type of reaction is sometimes called an inner salt. The conventional notation for the reaction is

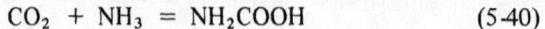

$$CO_2 + NH_3 = NH_2COOH \qquad (5\text{-}40)$$

The product shown in Eq. (5-40), carbamic acid, is known only in its salts; the net reaction is actually

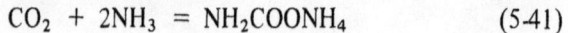

$$CO_2 + 2NH_3 = NH_2COONH_4 \qquad (5\text{-}41)$$

That is, ammonium carbamate is formed. This material often plugs screens in ammonia compressors, where it appears as damp, feathery crystals. Ammonium carbamate, if gently heated, forms urea

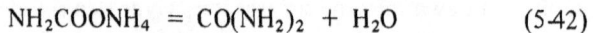

$$NH_2COONH_4 = CO(NH_2)_2 + H_2O \qquad (5\text{-}42)$$

Urea hydrolyzes in boiling water to ammonia and carbon dioxide.

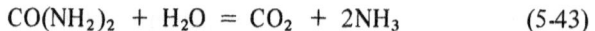

$$CO(NH_2)_2 + H_2O = CO_2 + 2NH_3 \qquad (5\text{-}43)$$

Thus, all the possible products of reaction between ammonia and carbon dioxide are thermally labile, so that in a hot steam condensate system no permanently stable neutralization products are formed. Protection is achieved entirely by decreasing the activity of the hydrogen ion, i.e., by elevation of pH. Because of its very low equivalent weight 0.25 ppm of ammonia is sufficient to raise the pH of pure water to 9.0. The distribution ratio of ammonia between steam and hot water, however, is about 10:1 so that enough must be added to the system to produce 2.5 ppm of ammonia in the steam to give a pH of 9.0 in the condensate.

A serious disadvantage of ammonia is that in the presence of oxygen at pH above 8.3 it attacks nonferrous metals—in particular brass and other alloys of copper. As mentioned in Section 5.3b of this chapter, the corrosive effect of oxygen on copper is greatly intensified in the presence of ammonia because of the stability of the cupric ammonia complex ion, $Cu(NH_3)_4^{++}$.

$$Cu + 4NH_3 + \tfrac{1}{2}O_2 = Cu(NH_3)_4^{++} + 2OH^- \qquad (5\text{-}18)$$

This reaction can cause severe corrosion in condensers, economizers, and feed water preheaters constructed of copper alloys. Further damage is done to boiler tubes by the soluble copper ammonia ion, which plates on steel, becomes a cathode, and causes the surrounding and underlying iron to dissolve.

$$Cu(NH_3)_4^{++} + Fe + 2OH^- = Cu + Fe(OH)_2 + 4NH_3 \qquad (5\text{-}44)$$

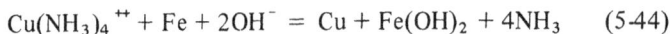

The ammonia returns to the steam system to function again as a transfer agent for copper. Similar ammonia complexes are formed with nickel and zinc ions.

b. Neutralizing Amines

Amines are derivatives of ammonia in which one or more hydrogen atoms have been replaced by an organic radical; they are sometimes called nitrogen bases. The neutralizing amines most commonly used to control corrosion in steam condensate systems are morpholine and cyclohexylamine. Hexylamine and benzylamine have been investigated and are often mentioned, but they are not used as commercial inhibitors. Recently diethylaminoethanol, 2-amino-2-methyl-1-propanol, and 2-dimethylamino-2-methyl-1-propanol have been described as useful volatile amines.[48]

Table 5.14 lists some physical properties that are significant in determining the most effective methods for using neutralizing amines or ammonia.

TABLE 5.14

Physical Properties of Neutralizing Amines

Chemical	Molecular weight	Boiling point (C)	pK_a	Partition coefficient (s/c)
Morpholine	87.12	128.9	8.36	0.4/1
Hexylamine	101.19	132.7	10.64	2/1 (est.)
Cyclohexylamine	99.19	134.5	10.83	3/1
Benzylamine	107.16	184.5	9.30	4/1
Ammonia	17.03	−33.4	9.26	10/1

Obrecht,[49] in an exemplary series of articles, has given a thorough quantitative discussion of the use and function of amines in preventing corrosion in steam condensates. He considers the partition coefficient, s/c, the ratio of the concentration of amine in the vapor phase to that in the liquid phase, to be the most significant factor in selecting neutralizing amines.

One of the more useful relationships for predicting properties of organic compounds is that of structural iteration, a term proposed by Branch and

Calvin,[50] of which structural homology is a special case. In general, it is found that in a homologous series if some physical property, e.g., melting point, boiling point, specific gravity, etc., is plotted against the number of carbon atoms in the corresponding compound a straight line is obtained. The series of amines under consideration unfortunately is not amenable to this method, as each amine belongs to a class of compound different from the others. Thus, benzylamine is a primary aromatic amine, cyclohexyl-amine is a primary amine of a cyclic paraffin, hexylamine is a primary alkyl amine, and morpholine is a six-membered heterocycle, specifically an imide oxide, or isoxazine.

Other properties that might be significant in determining the partition coefficient, s/c, are solubility in water, basic strength, vapor pressure, and heat of solution. Davison, et al.[51] have investigated the relationship between structure and solubility in water of many amines. They noted that substituents that increased the electron density on the nitrogen atom increased the solubility of the compound in water. Also, the longer the chain attached to the amino group, the lower the heat of solution, and the less soluble is the amine in water.

Before considering the basic strengths of the neutralizing amines more closely, Table 5.15, where the vapor pressures at various temperatures of the pure amines are given, should be examined. It is seen here that the partition coefficients are inversely proportional to the vapor pressures of the amines, an anomolous relationship.

TABLE 5.15

Vapor Pressure of Neutralizing Amines

Chemical	Ref.	Vapor pressure (mm Hg @ $t°C$)						Partition coefficient (s/c)
		20°	40°	60°	80°	100°	120°	
Morpholine	(52)	7.4	20	56	153	420	720	0.4/1
Hexylamine	(53)	7.2	20	54	135	275	530	2/1 (est.)
Cyclohexylamine	(54)	5.4	16	45	130	270	506	3/1
Benzylamine	(53)	0.5	2.1	6.7	19	44	100	4/1

The higher apparent volatility of cyclohexylamine as compared to that of morpholine has been attributed to the formation of an azeotrope of cyclo-hexylamine and water, that boils at 96.4 C. This is a dubious explanation, however, as the apparent volatility of benzylamine is still higher, even though it does not form an azeotrope.

Basicity in organic compounds nearly always arises from the presence of unshared electron pairs. Thus, amines produce an alkaline reaction in aqueous solution by functioning as an electron donor, or Lewis base, with-drawing hydrogen ions from water and leaving an excess of hydroxyl ions in the solution. The lower alkyl amines are stronger bases than ammonia because of the inductive effect[50] of methylene groups. These substituents increase the density of electrons on the nitrogen atom and thus its affinity for protons.

In Table 5.14 the values of pK_a for each amine and for ammonia are given. The modern practice in presenting tables of dissociation con-stants[55] is to list pK_a values for both acids and bases with the understand-ing that

$$pK_a = -\log_{10}K_a = 14 - pK_b \qquad (5\text{-}45)$$

For amines

$$K_a = \frac{(RNH_2)(H^+)}{(RNH_3{}^+)} \qquad (5\text{-}46)$$

The alkaline reaction obtained upon dissolving an amine in water can be represented thus:

$$RH_2N\colon + H^+OH^- = RH_2\overset{+}{N}\colon H + OH^- \qquad (5\text{-}47)$$

The basic dissociation constant for this proton transfer reaction is then

$$K_b = \frac{(RNH_3{}^+)(OH^-)}{(RNH_2)} \qquad (5\text{-}48)$$

The pH values in Table 5.16 are calculated from K_b by the following method.

Example: Calculate the pH of pure water containing 10 ppm of morpho·

TABLE 5.16

Effect of Amines on Alkalinity of Pure Water

Base	pK_a	pK_b	pH 1.0 ppm	pH 10 ppm	pH 100 ppm
Morpholine	8.36	5.64	8.68	9.18	9.71
Hexylamine	10.83	3.36	8.98	9.91	10.24
Cyclohexylamine	10.64	3.17	9.20	9.95	10.68
Benzylamine	9.30	4.70	9.35	9.95	10.48
Ammonia	9.26	4.74	9.39	9.97	10.50

line. Calling morpholine M, with $K_b = 2.3 \times 10^{-6}$, the following can be written:

$$K_b = \frac{(MH^+)(OH^-)}{(M)} = 2.3 \times 10^{-6}$$

Noting that $(MH^+) = (OH^-)$,

$$\frac{(OH^-)^2}{(M) - (OH^-)} = 2.3 \times 10^{-6}$$

Also, 10 ppm of morpholine is 0.010 g/l, or $0.010/87 = 1.15 \times 10^{-4}$ mole/l. Substituting and rearranging,

$$(OH^-)^2 + 2.3 \times 10^{-6}(OH^-) - 2.65 \times 10^{-10} = 0$$

Applying the quadratic formula,

$$(OH^-) = \frac{-2.3 \times 10^{-6} + \sqrt{(2.3 \times 10^{-6})^2 + 10.6 \times 10^{-10}}}{2}$$

$$= \frac{-2.3 \times 10^{-6} + \sqrt{10.55 \times 10^{-10}}}{2}$$

$$= 1.51 \times 10^{-5}$$

$$(H^+) = 10 \times 10^{-15}/1.51 \times 10^{-5}$$
$$= 6.62 \times 10^{-10}$$
$$pH = 10 - \log_{10}(6.62)$$
$$= 10 - 0.82$$
$$= 9.18$$

Comparing pK_b in Table 5.16 with s/c in Table 5.15 it is seen that the strongest base, cyclohexylamine, has a partition coefficient of 3, while the weakest base, morpholine, has a partition coefficient of 0.4; another anomolous relationship. The theoretical approach having failed completely, it now becomes necessary to approach this from a practical standpoint.

With very few exceptions, commercial formulations of neutralizing amines contain 15-30 percent of morpholine and 25-85 percent of cyclohexylamine, with from a few parts per million of water to 60 percent. On account of their partition coefficients, these two amines are especially effective for controlling corrosion by carbon dioxide. Morpholine dissolves readily in the first condensate that forms, thus giving protection to turbines and other equipment exposed to wet steam. It also has little tendency to escape through vents, or from deaerators. Morpholine alone is sometimes used in high-pressure systems having low make up and relatively short distribution. Cyclohexylamine is suited to low pressure systems with extensive steam distribution and long lines, but it escapes easily from vents, flash tanks, and deaerators. Blends of the two amines are recommended for industrial systems operated at several steam pressures.

Neutralizing amines are most conveniently stored in covered, stainless steel bulk tanks. Fed continuously by proportioning pumps, they can be injected directly into the steam header through stainless steel nozzles, but more often, as far as possible downstream from the deaerator to allow time for sodium sulfite to react with oxygen. If these chemicals are received in barrels, careful attention must be given to handling procedures. All neutralizing amines cause severe burns and are irritating to skin, eyes, and mucous membranes. Morpholine may cause liver and kidney damage. Cyclohexylamine can cause convulsions and has been known to produce cancer in some animals.

Jacklin[56] has investigated the thermal stability of a number of neutralizing amines by measuring the concentration of ammonia in steam generated at various temperatures and pressures. The results for morpholine and cyclohexylamine are given in Table 5.17. The concentration of each amine in the feed water is 34 ppm.

TABLE 5.17

Thermal Stability of Neutralizing Amines

Pressure (psi)	Temperature (F)	NH$_3$ in steam (ppm)	
		Morpholine	Cyclohexylamine
500	467	0.34	—
800	518	0.24	—
1500	596	0.49	0.31
2500	669	0.73	—

These tests were done with 100 percent make up, i.e., no condensate was returned. In actual practice the concentration of ammonia in steam treated with neutralizing amines is higher than indicated in Table 5.17.

In the following discussion a most important consideration in treatment with neutralizing amines is considered: dosage and cost. At the beginning of this section it was stated that some 80 percent of carbonate ions are converted to carbon dioxide in a steam generator. Eq. (3-2) shows that two moles of bicarbonate yield one mole each of carbonate and carbon dioxide or, ultimately, 1.8 moles of carbon dioxide. To calculate the amount of amine required to neutralize the potential carbon dioxide, it is necessary to determine the concentration of bicarbonate and carbonate ions in the feed water. This is done by measuring the P- and M-alkalinities and then referring to Table 5.4. The second condition (P $<$ M/2) being the most common in boiler feed water is used as an example. When P is less than M/2, carbonate is the equivalent of 2P ppm of CaCO$_3$ and bicarbonate is the equivalent of (M − 2P) ppm CaCO$_3$. These values are converted to ppm CO$_3^{--}$ and ppm HCO$_3^-$ as follows:

$$\text{ppm CO}_3^{--} = 2P \times 0.60$$

$$\text{ppm HCO}_3^- = (M - 2P) \times 1.22$$

Also 80 percent of CO_3^{--} and 90 percent of HCO_3^- are converted to CO_2. Finally, multiplying by the appropriate molecular weight ratios yields:

$$\text{ppm } CO_2 = [2P \times 0.60 \times 0.80 \times 44/60] +$$

$$[(M - 2P) \times 1.22 \times 0.90 \times 44/61]$$

$$= (0.79M - 0.88P) \tag{5-49}$$

In order to estimate the cost of treatment and to relate the unit costs of proprietary formulations to their neutralizing capacity, their equivalent weights are determined by the procedure in Chapter 8. In Table 5.18 are unit prices, basicities, and equivalent weights as bases of three commercial formulations of neutralizing amines.

TABLE 5.18

Costs and Strengths of Amine Formulations

Formulation	Unit price ($/lb)	Basicity (meq/g)	Equivalent weight (g/eq)
A	1.13	9.56	105
B	0.652	6.16	162
C	0.804	7.95	126

It is necessary to determine the relative costs of these three formulations when used to neutralize carbon dioxide. Amines react with carbon dioxide to form very weak carbamic acids:

$$RNH_2 + CO_2 = RNHCOOH \tag{5-50}$$

Eq. (5-50) shows that one mole (also one equivalent) of base neutralizes one mole of carbon dioxide. From Table 5.18 it is seen that 105 lb of Formulation A are needed to neutralize 44 lb of CO_2, or 2.39 lb of A/lb of CO_2. Table 5.19 summarizes results for each formulation.

TABLE 5.19

Cost-Effectiveness of Amine Formulations

Formulation	lb Formulation/lb CO_2	$/lb CO_2
A	2.39	2.70
B	3.68	2.40
C	2.86	2.30

Thus, Formulation C, though second highest in unit price, is the least expensive product to use. As a specific example, suppose that P = 4 and M = 10 in a feed water. The potential CO_2 is then ($10 \times 0.79 - 4 \times 0.88$), or 4.4 ppm. To neutralize this with Formulation C would cost 4.4×2.30, or $10.12/10^6$ lb of steam, which in most instances is prohibitively expensive. A much cheaper alternative is considered below.

c. Filming Amines

Primary alkyl amines with long straight chains form hydrophobic polar films on the surface of metal that physically block the discharge of hydrogen gas and metal ions. When properly applied, these film-forming amines offer significant advantages over neutralizing amines. Their dosage is independent of the concentration of carbon dioxide, they are effective at low concentration, and they protect metal surfaces from all dissolved gases. The best protection is afforded by octadecylamine, $C_{18}H_{37}NH_2$; amines with branched or unsaturated chains are less effective, as also are tertiary amines and mixtures.

Octadecylamine attaches itself to a metal surface by a chemisorption bond through the nitrogen atom, with the eighteen-membered carbon chain oriented perpendicular to the surface of the metal. Side chains interfere with this orientation. If three round pencils are held together in closed triangular packing so that each is tangent to the other two and the assembly is held vertically with the erasers against this page, a model of a unit of an amine protective film will be formed. The page represents the surface of the metal; the erasers are the nitrogen atoms; and the shafts of the pencils are the octadecyl carbon chains.

Fig. 5.2 is a schematic representation of a unit of monomolecular film in condensed close packing arrangement. Again, if this page is the surface

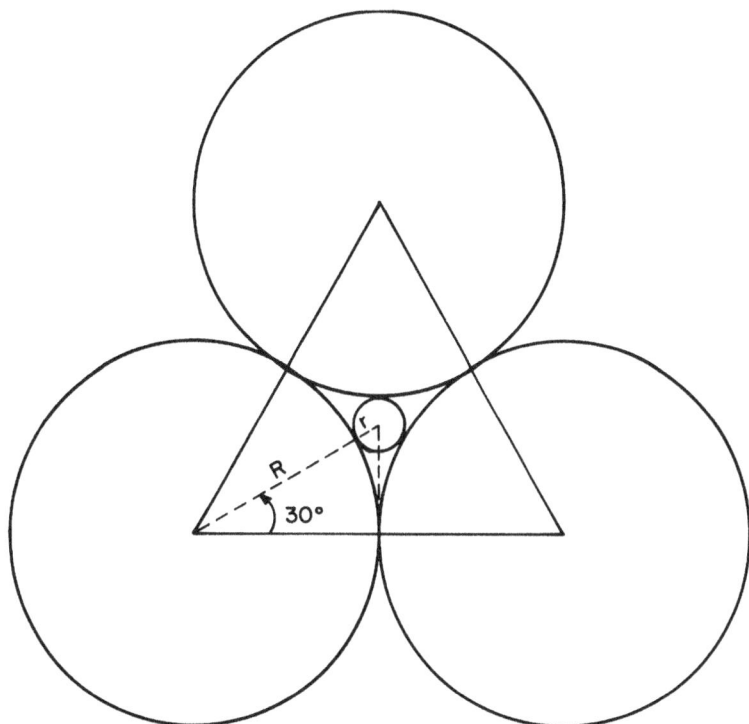

Fig. 5.2 Schematic top view of chemisorbed octadecylamine molecules.

of the metal, the reader is looking down on the ends of the carbon chains which are the large circles. It can be seen in Fig. 5.2 that

$$\cos 30° = R/(R + r)$$
$$0.866 = R/(R + r)$$
$$r = R(1 - 0.866)/0.866$$
$$= 0.155R$$

As the average cross-sectional area of polar alkyl radicals is about 20Å^2, R can be evaluated.

$$\pi R^2 = A_1 = 20$$
$$R^2 = 20/\pi$$
$$= 6.366$$
$$R = 2.52\text{Å}$$

Therefore,

$$r = 0.155 \times 2.52$$
$$= 0.39\text{Å}^2$$

and

$$A_s = 0.48\text{Å}^2$$

Thus, the area of the small circle (A_s) enclosed by the three mutually tangent large circles $(A_1 = 20\text{Å}^2)$ is $\approx \frac{1}{2}\text{Å}^2$. McClellan and Harnsberger[57] have reviewed the methods for measuring the cross-sectional area of molecules. The cross-sectional areas, in square Ångstroms, of several gas molecules of interest in steam-condensate systems are: H_2O, 12.5; O_2, 13.6; NH_3, 14.0; SO_2, 19; H_2S, 21; and CO_2, 21.8. Thus, these molecules are 25–45 times too large to enter the free space between the oriented carbon chains and therefore, cannot reach the surface of the metal.

The electronic theory of atomic structure postulates the existence of orbitals at various energy levels occupied by one electron, or by two electrons with opposed spins. The formation of bonds is explained by the assumption that incompleted orbitals in one atom can be filled by electrons from another atom. It was noted previously that in ammonia and amines the nitrogen atom possesses an unshared pair of electrons. Moreover, several elements in the first transition series contain incomplete $3d$ orbitals. Iron, in particular, has four $3d$ orbitals containing only one electron each. The possibility therefore exists for the formation of a chemisorption bond through interaction of the unshared electron pair of nitrogen and the incomplete orbitals of iron.

Trabanelli and Carassiti[58] have discussed the mechanism of corrosion inhibition by organic nitrogen compounds, and point out that the strength of an adsorption bond is determined by the electron density on the atom acting as the reactive center and by the polarizability of the function. Thus, increasing length of the hydrocarbon chain of aliphatic amines produces an increase in inhibitory efficiency that is attributed to the inductive effect of the methylene groups. The surface charge of the metal being inhibited also plays a role.

In addition to octadecylamine, hexadecylamine, $C_{16}H_{33}NH_2$, and the secondary amine dioctadecylamine, $(C_{18}H_{37})_2NH$, have been used to form polar films. As all of these compounds are virtually insoluble in water they are formulated as alkaline emulsions that disperse readily in warm condensate forming stable suspensions with a pH of about 9.5. The dilute emulsion is then pumped through a stainless steel injection nozzle into the boiler feed line or into the steam header. Continuous feed is advisable as the monomolecular film is not especially tenacious.

Ryznar and Kirkpatrick[59] have proposed the use of certain reaction products of aliphatic fatty acids with polyamines. A series of similar compounds is sold under the trade name Duomeens.* These have the general formula, $R\text{-}NH\text{-}(CH_2)_3\text{-}NH_2$, where R is an alkyl radical derived from coconut, soya, or tallow fatty acids.[60] The product made from beef tallow, for example, is a mixture of derivatives of myristic, palmitic, stearic, oleic, and linoleic acids; the diamine from oleic acid is called N-oleyl-1,3-propylenediamine.

Bregman[61] describes an interesting derivative of imidazoline, 1-ethyl-amino-2-octadecylimidazoline, which apparently is used in some commercial formulations as a film-forming amine.

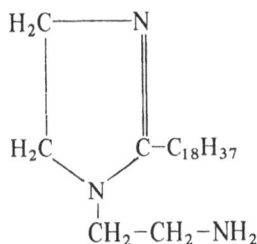

$$
\begin{array}{c}
H_2C \!\!-\!\!-\!\!-\!\! N \\
| \qquad\quad \| \\
H_2C \qquad C\text{-}C_{18}H_{37} \\
\diagdown \!\! N \diagup \\
| \\
CH_2\text{-}CH_2\text{-}NH_2
\end{array}
$$

As there is general agreement that optimum results are obtained with octadecylamine,[46,49,62] the remainder of the discussion is confined to this compound.

Octadecylamine is a waxy solid, insoluble in water, with a melting point of about 160 F. It steam-distills rapidly and completely, showing no tendency to concentrate in boilers. Wilkes, et al.[62] have measured the vapor pressure of octadecylamine and calculated its concentration in steam satur-

---

* Manufactured by Armour and Company, 100 S. Wacker Drive, Chicago, Ill. 60680.

ated with water and amine at various temperatures; these results are listed in Table 5.20.

TABLE 5.20

Vapor Pressure and Concentration of Octadecylamine in Steam

Temperature (C)	Vapor pressure (mm Hg)	Concentration in steam saturated with water and amine (ppm)
100	0.025	500
120	0.120	1,200
140	0.456	2,500
180	3.980	8,000
220	21.200	18,000
300	254.000	60,000
370	1197.000	115,000

These data show that octadecylamine is easily carried in steam and can separate only at low temperatures. Frequently, the acetate of octadecylamine is used. This compound should be fed to the boiler rather than to the steam header. In the latter, insoluble substituted acetamides are formed that subsequently deposit in condensate and feed water lines, as well as in strainers and steam traps.

The affinity for metal of the unshared electron pair on the nitrogen atom in filming amines causes the displacement of corrosion products and fouling of all types, which then collect in steam traps, on strainers, and at other restrictions. Because of this rummaging effect on corrosion products it is essential to begin treatment at low concentration, say 0.05 ppm of active amine, to avoid shutdowns for cleaning traps and strainers. Neutralizing amines do not cause these difficulties so can be started at the calculated dosage. Obrecht[49] suggests a maintenance dosage of 1-3 ppm of octadecylamine in industrial systems and 0.02–0.05 ppm in high-pressure utility systems. The inhibitor is lost to rust, oil, fouling deposits, and condensate drips. Therefore, when applying filming amines to a dirty system several months may elapse before maximum corrosion protection is realized. Oil, boiler water carryover, and process leaks strip the polar film from the surface of metal.

d. Combinations of Amines

Proprietary mixtures of filming and neutralizing amines are widely marketed. Denman[63] has proposed, for example, an aqueous emulsion of octadecylamine, cyclohexylamine, and a nonionic surfactant, such as one of the polyoxyalkylene glycols. Commercial preparations are designed to form stable emulsions when diluted with water. Dilutions are usually recommended if the amines are to be injected into the steam system, as the flashing water aids distribution. A specially designed nozzle is required to avoid steam collapse. Typical commercial blends contain 3-15 percent filming amine, 15-20 percent neutralizing amine, and 60-80 percent water, along with 1-2 percent of an emulsifying agent. They may contain either morpholine, cyclohexylamine, or both.

The idea of a combination of neutralizing and filming amines appeals strongly to the imagination for it implies a doubling of beneficial effect. Under scrutiny, however, the rationale of this treatment is elusive, to say the least. Let us consider, for instance, an actual proprietary blend containing 15 percent filming amine, 20 percent neutralizing amine, and 65 percent water. Its price is $0.60/lb, and the dosage recommended by the vendor is 3 ppm. Its equivalent weight is found to be 452 by the method in Chapter 8, so, as explained in Section 5.8b, 452 lb are needed to neutralize 44 lb of carbon dioxide, i.e., 10.3 lb/lb CO_2. If this blend is added to a feed water containing, say, 4.4 ppm potential CO_2, 4.4 × 10.3 = 45.5 lb (at a cost of $27)/$10^6$ lb of steam generated must be used. As 15 percent of the blend is filming amine, its concentration in the steam will be 45.5 × 0.15 = 6.8 ppm, which is 2-7 times too much for industrial systems and 130-350 times too much for utilities. On the other hand, if the recommended dosage of 3 ppm is used, the result will be 0.45 ppm of filming amine, but only 0.6 ppm of neutralizing amine. Thus, in this case a compromise dose of filming amine along with 6 or 7 percent of the amount of base required to neutralize carbon dioxide is being added. It is sometimes claimed that filming amines bond to metal more tightly if the pH is raised above 6. If this is indeed the case, ammonia would be much more economical than neutralizing amines in mixed formulations.

As a final word, users of amines should be aware that objectionable tastes and odors can be imparted to food being processed by steam and also that nitrogen bases poison some types of catalysts used for refining petroleum.

REFERENCES

(1) Green, J. 1950. Reversion of molecularly dehydrated sodium phosphates. *Ind. Eng. Chem.* 42:1542.

(2) Jenkins, D., Ferguson, J. F., and Menar, A. B. 1971. Chemical processes for phosphate removal. *Water Research* 5:369.

(3) Hall, R. E., Smith, G. W., Jackson, H. A., Robb, J. A., Karch, H. S., and Hartzell, E. A. 1927. *A physico-chemical study of scale formation and boiler-water conditioning.* Bulletin 24, Carnegie Institute of Technology, Pittsburgh.

(4) Bell, W. E. 1965. Chelation chemistry—its importance to water treatment and chemical cleaning. *Mater. Protect.* 4 (2):78.

(5) Meites, L. 1963. *Handbook of analytical chemistry.* New York: McGraw-Hill. pp. 1–45, Table 1–19.

(6) Edwards, J. C. and Rozas, E. A. 1961. Boiler-scale prevention with EDTA chelating agents. *Proc. Amer. Power Conf.* 23:575.

(7) Merriman, W. R. 1964. Continuous boiler treatment with EDTA: a progress report. *Proc. Intl. Water Conf., Eng. Soc. of W. Pa.* 25:157.

(8) Jacklin, C. 1965. Chelating agents for boiler treatment—research and actual use. *Proc. Amer. Power Conf.* 27:807.

(9) Walker, J. L. and Stephens, J. R. 1973. A comparative study of chelating agents: their ability to prevent deposits in industrial boilers. *Proc. Intl. Water Conf.* 34th Proceedings, 561.

(10) Lux, J. A. 1965. Chelating agents for boiler treatment—a manufacturer's viewpoint. *Proc. Amer. Power Conf.* 27:817.

(11) Schantz, J. 1966. Chelating agent for boiler treatment. *Hawaii Sugar Technol.* Rep. 25:100.

(12) Noll, D. E. 1958. Limitations on chemical means of controlling corrosion in boilers. *Corrosion* 14:541t.

(13) Pfeil, L. B. 1929. The oxidation of iron and steel at high temperatures. *Iron and Steel Inst. J.* 119:501.

(14) Douglas, D. L. and Zyzes, C. F. 1957. The corrosion of iron in high temperature water—part 2. *Corrosion* 13:433t.

(15) Hamer, P., Jackson, J., and Thurston, E. F. 1961. *Industrial water treatment practice.* London: Butterworths.

(16) Pye, D. J. 1947. Chemical fixation of oxygen. *J. Amer. Water Works Assoc.* 39:1121.

(17) Alexander, R. C. and Rummel, J. K. 1950. The quality of steam condensate as related to sodium sulphite in boiler water. *Trans. ASME* 72:519.

(18) Fiss, E. C. 1955. The use of hydrazine to prevent oxygen cor-

rosion. *Southern Power and Ind.* 73 (12):46.
(19) Everitt, G. E. 1962. Industrial efficacy of hydrazine as an oxygen scavenger. *Chem. Ind.* (London) 40:609.
(20) Stones, W. F. 1957. The use of hydrazine as an oxygen scavenger in h. p. boilers. *Chem. and Ind.* (London) 35:120.
(21) Uhlig, H. H. 1967. *Corrosion and corrosion control.* New York: John Wiley and Sons.
(22) Purcell, T. E. and Whirl, S. F. 1943. Protection against caustic embrittlement by coordinated phosphate-pH control. *Trans. Electrochem. Soc.* 83:343.
(23) Parr, S. and Straub, F. 1926. The cause and prevention of embrittlement of boiler plate. *Proc. Amer. Soc. Testing Materials* 26 Part II:52.
(24) Blackburn, T. R. 1969. *Equilibrium: a chemistry of solutions.* New York: Holt, Rinehart, and Winston.
(25) Noll, D. E. 1964. Factors that determine treatment for high-pressure boilers. *Proc. Amer. Power Conf.* 26:753.
(26) Marcy, V. M. and Halstead, S. L. Jan., 1964. Improved basis for coordinated phosphate-pH control of boiler water. *Combustion* 35:45.
(27) Ravich, M. I. and Shcherbakova, L. A. 1955. Nature of the solid phase which crystallizes from aqueous solutions of trisodium orthophosphate at high temperatures. *Izvest. Sektora Fiz-Khim. Anal., Inst. Obshekei i Neorg. Khim., Akad. Nauk S. S. S. R.* 26:248; C.A., 50: 3055i (1956).
(28) Jacklin, C. and Brower, S. R. *Correlation of silica carryover and solubility studies.* ASME Paper No. 51-A-91.
(29) Coulter, E. F., Pirsh, E. A., and Wagner, E. J., Jr. 1956. Selective silica carry-over in steam. *Trans. ASME* 78:869.
(30) Straub, F. C. and Gravowski, H. A. 1945. Silica deposition in steam trubines. *Trans. ASME* 67:309.
(31) Splittgerber, A. 1941. The volatility of silicic acid. *Archiv. für Wärmewirtschaft* 22:66.
(32) Latimer, W. M. 1952. *The oxidation states of the elements and their potentials in aqueous solutions.* 2nd ed., Englewood Cliffs: Prentice-Hall.
(33) Garrels, R. M. and Christ, C. L. 1965. *Solutions, minerals, and equilibria.* New York: Harper and Row.
(34) Dwyer, J. L. and Frith, C. F. 1967. New analytical techniques for determination of colloidal contamination in high purity steam generating systems. *Proc. Amer. Power Conf.* 29:778.
(35) Gray, J. A. 1957. Boiler water treatment: a formula for the

control of sludge and scale in internal (carbonate) treatment. *J. Inst. Fuel* 66:577.

(36) Denman, W. L. and Salutsky, M. C. Sept., 1968. Boiler scale control. *Power* 112 (9):80.

(37) Holmes, J. A. and Jacklin, C. Jan., 1944. Experimental studies of boiler scale at 800 psi. *Combustion* 15 (7):35.

(38) Sommerauer, A., Sussman, D. L., and Stumm, W. 1968. The role of complex formation in the flocculation of negatively charged sols with anionic polyelectrolytes. *Kolloid Zeitschrift und Zeitschrift für Polymere* 225:147.

(39) Buehrer, J. F. and Reitemeier, R. F. 1940. The inhibiting action of minute amounts of sodium hexametaphosphate on the precipitation of calcium carbonate from ammoniacal solutions, II. *J. Phys. Chem.* 44:552.

(40) Zeleny, R. A. and Vithani, K. 1963. The role of organic additives in preventing scale formation on heating surfaces. *Combustion* 34 (8):47.

(41) Jacoby, A. L. and Bischmann, L. C. 1948. Steam bubble formation. *Ind. Eng. Chem.* 40:1360.

(42) Gunderson, L. O. and Denman, W. L. 1948. Polyamide foam inhibitors. *Ind. Eng. Chem.* 40:1363.

(43) Fordyce, R., Lovell, E. L., and Hibbert, H. 1939. Studies on reactions relating to carbohydrates and polysaccharides. LVI. The synthesis of the higher polyoxyethylene glycols. *J. Amer. Chem. Soc.* 61:1905.

(44) Hall, R. E., Smith, G. W., Jackson, H. A., Robb, J. A., Karch, H. S., and Hartzell, E. A. 1927. *A physico-chemical study of scale formation and boiler-water conditioning.* Bulletin 24. Mining and Metallurgical Investigations. Carnegie Institute of Technology, Pittsburgh, Pa.

(45) Schenk, J. E. and Weber, W. J. Jr. 1968. Chemical interactions of dissolved silica with iron (II) and (III). *J. Amer. Water Works Assoc.* 60:199.

(46) Maguire, J. J. 1954. After boiler corrosion. *Ind. Eng. Chem.* 46:994.

(47) Luder, W. F. and Zuffanti, S. 1946. *The electronic theory of acids and bases.* New York: John Wiley and Sons.

(48) Cuisia, D. G. May, 1977. Experimental determination of the volatility of condensate corrosion inhibitors. *Materials Performance.* 16 (5):21.

(49) Obrecht, M. F. 1964. Cause and cure of corrosion in steam-condensate cycles - I. *Effluent Water Treat. J.* 4 (5):223; II, ibid. 1964. 4 (6):279.; III, ibid. 1964. 4 (7):320; IV, ibid.

1964. 4 (8):373.

(50) Branch, G. E. K. and Calvin, M. 1944. *The theory of organic chemistry.* New York: Prentice-Hall.

(51) Davison, R. R., Smith, W. H., Jr., and Hood, D. W. 1960. Structure and amine-water solubility in desalination by solvent extraction. *J. Chem. Eng. Data* 5:420.

(52) Landolt-Bornstein Zahlenwerte und Funktionen, 1963. Sechste Auflage, Vol. 2a: 139. Herausgegeben von Heinz Borchers und Ernst Schmidt. Berlin: Springer-Verlag.

(53) Jordan, T. E. 1954. *Vapor pressure of organic compounds.* New York: Interscience.

(54) Novák, J., Matous, J., and Pick, J. 1960. The vapor pressure of cyclohexanol and cyclohexylamine. *Collection Czechoslov. Chem. Communs.* 25:583; C. A. 54:16067i.

(55) Perrin, D. D. 1965. *Dissociation constants of organic bases in aqueous solution.* London: Butterworths.

(56) Jacklin, C. 1955. Experimental boiler studies of the breakdown of amines. *Trans. Amer. Soc. Mech. Engrs.* 77:449.

(57) McClellan, A. L. and Harnsberger, H. F. 1967. Cross-sectional areas of molecules adsorbed on solid surfaces. *J. Coll. and Interface Sci.* 23:577.

(58) Trabanelli, G. and Carassiti, V. 1970. Mechanism and phenomenology of organic inhibitors. Chapter 3 of *Advances in corrosion science and technology.* Vol. 1. ed. by Fontana, M. G. and Staehle, R. W. New York-London: Plenum Press.

(59) Ryzner, J. W. and Kirkpatrick, W. H. Nov. 20, 1956. *Inhibition of corrosion in steam-condensate lines.* U.S. Patent No. 2,771,417.

(60) Zimmermann, O. T. and Lavine, I. 1956. *Handbook of material trade names.* Supplement I. New Hampshire: Industrial Research Services.

(61) Bregman, J. J. 1963. *Corrosion inhibitors.* New York: Macmillan.

(62) Wilkes, J. F., Denman, W. L. and Obrecht, M. F. 1955. Filming amines: use and misuse in power plant water-steam cycles. *Proc. Amer. Power Conf.* 17:527.

(63) Denman, W. L. April 14, 1969. Corrosion inhibiting composition and method. U.S. Patent No. 2,882,171.

Chapter 6.

Operating Procedures

Having described a number of physical and chemical methods for preparing water to be transformed into steam, it is important to now examine some of the practical aspects of operating boilers and water-treating plants. The objectives in treating make up water for boilers, of course, are to minimize corrosion, to inhibit the formation of scales and deposits, and to produce clean steam of high quality safely and efficiently.

6.1   TYPES OF BOILERS

The quality and condition required of treated water is determined to a large extent by the type of boiler in which the water is to be used and by the operating pressure. Accordingly, it is appropriate to consider several possible applications.

a.   Stationary Field-Erected Boilers

Central power stations and numerous industrial plants produce steam from large generators assembled at the building site. Typically, these are water-tube boilers with two to four drums, fired by coal, fuel oil, or natural gas, with water tubes shielding the sides, top, and bottom of the furnace to cool the refractory. As a rule, utility steam generators have a large radiant surface that includes the superheater, and a relatively small convective surface. In order to improve efficiency, boilers in power plants are equipped with economizers, which transfer sensible heat from flue gas to the feed water just before the latter enters the steam drum. In addition, an air preheater is installed between the economizer and the stack to heat combustion air. Before being introduced into the furnace with the fuel, air is forced by a fan through the heater, where it is warmed by heat reclaimed

120

from flue gas. Heated combustion air provides significant fuel savings, makes combustion more nearly complete, reduces the concentration of sulfur trioxide in flue gas, lowers the stack temperature to about 200 F, and raises the temperature of the furnace.

In plants that generate electricity, the bulk of the feed water is returned condensate supplemented by 1 percent, or less, of treated water make up. Operating pressures range from a minimum of 600 psi to as much as 5000 psi.

b. Stationary Packaged Boilers

The modern approach to designing boilers concentrates on using the minimum of steel to achieve the maximum of heat flux; this necessarily demands a high ratio of heating surface to water capacity. This trend has led to the development of the packaged steam generator, shop-assembled, and shipped complete with fuel burners, refractory, external insulation, soot-blower, draft fan, piping and wiring, together with the necessary gauges, safety valves, and control panel. Economizer, air heater, and combustion controls are also included, if desired. The overall dimensions and weight of these units are restricted by railroad shipping clearances, so they are extremely compact; typically, 11 ft wide, 15 ft high, and 25-30 ft long. Combustion efficiency is high, heating is extremely rapid from a cold start on account of the small volume of contained water, and heat flux is enormous—up to 250,000 Btu/h/ft^2 compared to 50,000 Btu/h/ft^2 in boilers designed 30-40 years ago. Operating pressures range from less than 250 psi to 900 psi; a few packaged units have been built to operate at 2000 psi. Because of the intense heat flux in these units, feed water quality must be substantially better than for a field-erected boiler operated at the same pressure.

c. Supercritical Boilers

As pressure increases in a steam generator the density of steam increases while that of water decreases. At the critical pressure, 3203.6 psi, the densities become equal, and there is no phase separation of water and steam. Within a supercritical boiler the transition from one to the other occurs at a point where the specific volume and enthalpy of the supercritical fluid undergo a rapid change with only a small change in temperature.[1] Throughout the heat absorption process at the critical pressure, however, a fluid state exists without the formation of bubbles or separation into two phases.

A supercritical boiler has no drum: it is, in effect, a single tube wherein the transition from water to steam occurs on the convection surfaces of the boiler and is regulated by matching the flow of water to that of steam. As there is no water level, the usual method of controlling feed water rate cannot be used. Furthermore, as there is no separation of phase, blowdown is also impossible. Because of exceedingly high temperatures and heat flux in these steam generators, only minute concentrations of dissolved or suspended solids can be tolerated, and the possibilities of internal chemical treatment are strictly limited. A few boilers have been built to operate at 5000 psi, but modern designers prefer to limit pressure to 3500 psi at the turbine throttle with a maximum temperature of 1050 F.[2]

d. Marine Boilers

The main boilers on ships provide steam for propulsion and with a few exceptions are fired with fuel oil. On large tankers, auxiliary steam generators of the shell and tube type are sometimes installed to supply steam for cargo and fuel oil heating coils, deck machinery, stripping pumps, and heating accommodations. Alternatively, 150-psi desuperheated or let-down steam from the main boilers is used for these services. Propulsion boilers operate at 900–1200 psi, using condensate combined with a small amount of make up water distilled from sea water in an evaporator. Marine boilers are usually fitted with economizers, but because of severely limited space, air heaters are omitted. As the concentrations of alkalinity and dissolved solids are low in evaporated feed water, the coordinated phosphate-pH treatment is especially suited to marine boilers. This subject is discussed in more detail further on.

e. Waste Heat Boilers

The term waste heat may be applied to fuels with little heat content— i.e., less than 5000 Btu/lb—such as bark, wood refuse, bagasse, coffee grounds and hulls, rice hulls, corn cobs, and black liquor from wood pulping plants.[3] More commonly, in industry, the term describes heat recovered, mostly by convection, in a nonfired steam generator, from hot gases or process streams produced in some manufacturing operation. In this arrangement a hot process stream flows through tubes surrounded by a shell containing water that is thus converted to steam, usually, though not invariably, at a pressure less than 150 psi. The advent of the fluid catalytic cracking process 25 years ago lead to the introduction of CO boilers[4] in which carbon monoxide from the catalyst regenerator is mixed with supplemental fuel and burned in a furnace to produce 800-psi process steam. This too is a recovery of otherwise wasted heat; the carbon monoxide leaves the regenerator at about 1000 F.

6.2 WATER TREATMENT PROGRAMS

In previous chapters the application of individual chemicals and their functions in treating water were discussed. Combinations of chemicals and sequences of treatments used to prepare water for the generation of steam are now considered.

Although water-treating formulations are marketed under a multitude of trade names and numbers, there are only a few fundamental methods of treating water for boilers. Table 6.1 is a summary of chemical treatments that have proven satisfactory for use in boilers at particular pressures. In preparing this table care has been taken to ensure that the recommended treatments are chemically consistent. Thus, lignins and tannins are known to char above 500 F,[5] so they are not appropriate for boilers operating above 600 psi. On the other hand, polyacrylates and polymethacrylates begin to char at ≈ 700 F, so there is no limitation on their use because of thermal considerations. Chelant treatments have been recommended for boilers operating between 150 and 1500 psi, but here 600 to 900 psi is suggested. Under 600 psi the appropriate methods of softening often do not reduce the total hardness of the make up water enough to make a chelant treatment economically feasible. Also, at 900 psi EDTA is thermally unstable, while at the same pressure NTA loses much of its chelating activity.[6] Polyoxyalkylene glycols function best as antifoams when the concentration of dissolved solids in boiler water is < 2000 ppm, whereas polyamides are much more tolerant of high concentrations of suspended solids. Consequently, the latter are limited to a maximum of 750 psi and the former are suggested between 750 and 1500 psi. These and similar considerations were taken into account in the preparation of Table 6.1.

At pressures below 500 psi it is usually unnecessary to make special provision for the removal of silica, for in plants where lime-soda softening is used the concentration of silica is significantly reduced in the softening process. For raw waters containing a preponderance of magnesium, how-

TABLE 6.1

Appropriate Treatments at Various Pressures

Pressure	Silica removal	Softening	Internal treatments			
			Antifoam	Antiscalant	Oxygen scavenger	Conditioner
100	—	Cold Lime-Soda	Polyamides	Sodium Carbonate	Sodium Sulfite	Lignin, Tannin
200	—	Hot Lime-Soda		Phosphate		
300	—	Hot Lime-Soda & Phosphate				Polyacrylate
500	—					
600	Lime-Magnesium Oxide	Cation Exchange				
750			Polyoxyalkylene glycols	Phosphate or chelant		
900	Demineralizing	Demineralizing				
1000				Phosphate-pH*	Hydrazine	
1500		Demineralizing & polishing		Phosphate-pH**		
2500						
3200+			—	—		—

* Coordinated treatment
** Congruent treatment

ever, lime-soda softening is so inefficient that magnesium salts may deposit in feedlines, filters, and economizers, or form sticky deposits of magnesium phosphate and sludges of magnesium hydroxide within boilers. This fouling can be avoided by adding sodium metasilicate to the softener, thereby precipitating magnesium silicate. Sometimes silicate is added directly to the boiler, but obviously this is less desirable, as it increases the concentration of suspended solids in the boiler water.

At pressures between 150 and 850 psi the so-called caustic reserve method of treatment is the oldest, least expensive, and still most widely used program for conditioning boiler water. Filtered, softened, deaerated water is used for make up. Phosphate is added to the boiler to precipitate residual alkaline earth elements, and catalyzed sulfite is used as an oxygen scavenger, supplemented by an antifoam and by natural or synthetic polymers added to disperse sludge. Free alkali is formed in the boiler by the thermal decomposition of carbonates and bicarbonates, producing a pH of 11.0–11.7. Advantages of the caustic reserve method of treatment are:

1. Its effectiveness.
2. The cost is low, as basic chemicals are used rather than expensive proprietary formulations.
3. Close control of chemical concentrations is not critical.
4. Chemical determinations can be made rapidly and do not require a great deal of skill.
5. Gasket and roll leaks are rare because of the self-sealing properties of sodium phosphate.

Disadvantages are:

1. The formation of finely divided precipitates ($0.5-5\mu$), which cause foaming.
2. The precipitation of excess sodium phosphate as a sludge as result of the free-and-easy way it is customarily added by boiler operators.

Sodium nitrate is often added to boiler water to prevent stress corrosion cracking, although the chemical is unnecessary in welded boilers and can actually be harmful at higher pressures (900–1500 psi). There are in service, however, a large number of low-pressure boilers with lap-riveted or butt-and-strap-riveted drums. These are prone to leak steam and water, leading to the local concentration of sodium hydroxide and the possibility

of caustic embrittlement; these steam generators (generally operated at
< 200 psi) should be treated with sodium nitrate. At higher pressures and
temperatures nitrate is successively reduced by magnetite to nitrite, hyponitrite, hydroxylamine, nitrous oxide, and ammonia. Also, at high temperature and high alkalinity nitrate thermally decomposes to nitrite and oxygen.

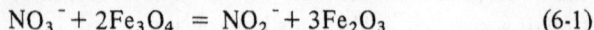

$$NO_3^- + 2Fe_3O_4 = NO_2^- + 3Fe_2O_3 \qquad (6\text{-}1)$$

$$2NO_2^- + Fe_3O_4 = N_2O_2^{--} + 6Fe_2O_3 \qquad (6\text{-}2)$$

$$4H_2O + N_2O_2^{--} + 4Fe_3O_4 = 2NH_2OH + 6Fe_2O_3 + 2OH^- \quad (6\text{-}3)$$

$$4NH_2OH = N_2O + 2NH_3 + 3H_2O \qquad (6\text{-}4)$$

It is essential when using nitrate in boilers ranging in pressure from 900-
1500 psi to be aware that each mole of nitrate produces one mole of hydroxyl ion and one-half mole of ammonia while oxidizing six moles of
magnetite, as shown by the net results of the above four equations.

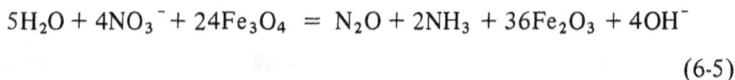

$$5H_2O + 4NO_3^- + 24Fe_3O_4 = N_2O + 2NH_3 + 36Fe_2O_3 + 4OH^-$$

$$(6\text{-}5)$$

In addition to the generation of free alkali, the concentration of ammonia
in the steam is greatly increased.

Currently, there is much interest and promotion of chelants as antiscalants for the internal treatment of boilers. It is alleged that the boiler
operator can expect cleaner boilers and higher rates of heat transfer with
chelant programs than are obtained with conventional phosphate treatments; under ideal conditions this is probably true. One should realize,
however, that over-treatment with chelants increases corrosion rates while
under-treatment produces troublesome scales. Deposits formed as a result
of insufficient treatment, e.g., pump failures, neglecting to prepare chelant
mixes, improper quantities in the mix, erroneous test results, are much
harder, more insoluble, and more adherent than those formed under phosphate treatment. For this reason, it is advocated by some that phosphate
be used with chelant to serve as a reserve of antiscalant in case of softener
leakage or the enforced use of hard water in emergencies. The work of
Walker and Stephens,[7] however, shows that if boiler water contains phosphate, hydroxyapatite, $Ca_5(OH)(PO_4)_3$, preferentially precipitates in the
presence of an excess of either or both EDTA or NTA, on account of the
exceedingly small solubility product ($K_{sp} = 3 \times 10^{-62}$) of this basic salt.

Thus, there is no point in using chelants at all in the presence of phosphate. Similarly, very little magnesium is chelated when silica is present at concentrations above 80 ppm, which suggests that a chelant treatment is not appropriate for boilers operating at < 500 psi. The solubility product of serpentine, $Mg_3(OH)_2(SiO_3)_2$, was empirically determined by Walker and Stephens[7] to be $K_{sp} = 3 \times 10^{-35}$.

In Table 6.2 are listed the formation constants (often called stability constants) and their logarithms of several chelonates of sodium (ethylene-dinitrilo)tetraacetate (EDTA) and sodium nitrilotriacetate (NTA), the two chelants in commercial use as antiscalants.[8] The formation constant is a measure of the forces holding a metal ion and a chelant together. Thus, the larger the formation constant, the more firmly bound the metal ion.

TABLE 6.2

Formation Constants of Selected Chelonates

Chelonate	K_f	pK_f
Ca EDTA	5.0×10^{10}	10.7
Mg EDTA	5.0×10^{8}	8.7
Fe EDTA (ous)	2.5×10^{14}	14.4
Fe EDTA (ic)	1.2×10^{25}	25.1
Co EDTA	2.0×10^{16}	16.3
Ca NTA	1.6×10^{8}	8.2
Mg NTA	1.0×10^{7}	7.0
Fe NTA (ous)	6.3×10^{8}	8.8
Fe NTA (ic)	7.9×10^{15}	15.9
Co NTA	4.0×10^{10}	10.6

Although the formation constants of EDTA are larger than those of NTA, the latter is thermally more stable and, thus, would seem to be the chelant of choice at intermediate pressures, i.e., 600 to 900 psi. Despite continual published allusions to the chelation of iron in boilers, it is obvious from the formation constants of the iron chelonates of EDTA and NTA and the solubility product of ferric hydroxide ($K_{sp} = 6 \times 10^{-38}$), that above pH 10 the chelation of ferric iron is impossible. For this reason, Fe_2O_3 and Fe_3O_4 are essentially unaffected by chelants.

Venezky,[9,10] in studying the thermal decomposition of EDTA and NTA, using the methods of nuclear magnetic resonance, found that EDTA decomposes to iminodiacetate and N-hydroxyethyliminodiacetate, the decomposition being accelerated by oxygen. Both of these degradation products are chelants, but they are much weaker than either EDTA or NTA.

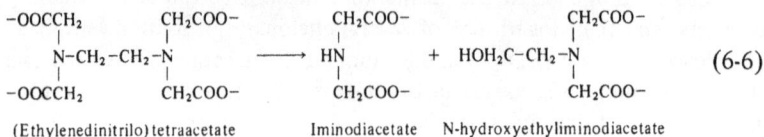

$$
\begin{array}{ccccc}
-OOCCH_2 & CH_2COO- & & CH_2COO- & CH_2COO- \\
| & | & & | & | \\
N-CH_2-CH_2-N & \longrightarrow & HN & + & HOH_2C-CH_2-N \\
| & | & & | & | \\
-OOCCH_2 & CH_2COO- & & CH_2COO- & CH_2COO-
\end{array} \qquad (6\text{-}6)
$$

(Ethylenedinitrilo)tetraacetate   Iminodiacetate   N-hydroxyethyliminodiacetate

Regardless of the chelants used and the claims made by vendors for particular formulations or combinations of products, it is essential that a polymer be included as an integral part of the program.[10a] A polymer is needed to disperse insoluble salts that form as a result of fluctuations in the hardness of feed water, failures of the chemical injection pump, or erroneous analyses. Suitable polymers for pressures in the range of 600–900 psi include sodium polyacrylate, sodium polymethacrylate, and carboxymethylcellulose. The latter chars at about 600 F, so should not be used above 1500 psi.

Instances of severe corrosion have been noted in steam drums into which concentrated chelant had been injected.[11] This damage often takes the form of thinning at the water line, with other attack concentrated in areas of the drum where the velocity of flow is high. Wastage is also enhanced at hot spots and in the upper sections of the risers; usually, chelant corrosion is uniformly dark and smooth and sometimes is not at all evident to the eye. To avoid this hazard, chelant solution should be fed continuously through a Type 304 stainless steel injection nozzle to the feed water. The point of injection should be as far as possible downstream from the deaerator if catalyzed sodium sulfite is used as an oxygen scavenger. Under no circumstances should these chemicals be mixed together, for chelants inactivate the cobalt catalyst (see Table 6.2), preventing the rapid reduction of oxygen. The latter enhances corrosive attack by NTA, hastens the thermal degradation of EDTA, and causes pitting in steel equipment, particularly economizers. If make up water is softened by cation exchange it is advisable to add neutralizing amines to the feed water to neutralize carbonic acid in the condensate. These compounds, being highly alkaline, should not be mixed with sodium sulfite, because of the possibility of precipitating $Co(OH)_2$; for the same reason they should not be added to the feed water near the deaerator outlet. Neutralizing amines should be injected into the suction side of the boiler feed pump, as far as possible downstream from the deaerator to allow the maximum time for sulfite to reduce oxygen. These important chemical considerations should never be ignored for the sake of mechanical expedience.

TABLE 6.3

Chelant Demand of Feed Water Hardness

	Disodium EDTA	Trisodium NTA
Formula Weight	372	257
ppm Chelant/ppm $CaCO_3$	3.7	2.6
ppm Chelant/ppm Ca	9.3	6.4
ppm Chelant/ppm Mg	15.5	10.7

In Table 6.3 the stoichiometric ratios of chelants to hardness are shown. In principle, chelant is continuously proportioned to the total hardness in the feed water by means of a chemical injection pump. For example, if the average concentration of hardness in a feed water is 0.5 ppm $CaCO_3$ (about the maximum permissible for a chelant program to be economically feasible), it would be necessary to maintain 1.9 ppm of EDTA, or 1.3 ppm of NTA, continuously in the feed. This leads to the concept of free, combined, and total chelant. Combined chelant is that portion of the total that exists in the form of chelonates of calcium and magnesium ion. The concentration of chelant present in excess of that combined with hardness is free (or residual) chelant. The sum of these two concentrations is, of course, the total chelant. The question of how much chelant is too much is moot, but at 900 psi 20 ppm of EDTA, or 30 ppm of NTA are reasonably conservative maximum concentrations of chelant. Practical difficulties arise, however, that should be appreciated before attempting to establish precise control of a chelant program. One of these is that it is impracticable in a boiler plant to measure accurately concentrations of total hardness between 0 and 1 ppm $CaCO_3$, even though the proper rate of addition of chelant depends upon values that lie in this range. Suppose, for instance, that a boiler operator determines that the concentration of total hardness in the feed water is 0.5 ppm $CaCO_3$. If EDTA is being used, by referring to Table 6.3 it is seen that the rate of addition should be 1.9 ppm of EDTA. If the blowdown rate is, say, 5 percent, the blowdown ratio is 20/1, and the concentration of combined EDTA in the boiler water is 38 ppm, with no residual chelant. Now let us suppose that the operator's result is erroneous and that the actual hardness is 0.1 ppm $CaCO_3$ in the feed water. In this case, the combined EDTA is 0.1 × 3.7 × 20, or 7.4 ppm, and the residual EDTA in the boiler water is (38 − 7.4) = 30.6 ppm, an excessive concentration.

When treating feed water containing 1–5 percent of make up, it is necessary to supplement chelant with sodium hydroxide in order to restrain the volatilization of silica and to prevent the deposition of aluminum oxide scale on boiler tubes. In the range of pressure under consideration (600–900 psi) the free alkalinity should be kept in the range of 60-120 ppm of NaOH. Equivalent alkalinities are: $(2P - M) = 75$-150 ppm $CaCO_3$ and B-reading = 3.6-7.2 ml.

Lorenc and Bermer[12] have disclosed a method for controlling scale with a combination of NTA and a phosphonate. In this process the concentration of total chelant in the boiler water is limited to 8-10 ppm, and the phosphonate (at about 1.5 ppm) is evidently depended upon to inhibit the formation of scales, as discussed by Ralston.[13] Obviously, though, if the chelant is over- or underfed, corrosion or scaling inevitably ensues. Polymers, antifoam, and hydrazine are also included in this proprietary method of treatment.

Walker and Stephens[7] have suggested a chelant treatment based on an equimolar mixture of EDTA and NTA that takes advantage of EDTA's capacity to form more stable complexes than NTA that also are thermally more stable than free EDTA. With this scheme it is presumed that the EDTA is combined and part of the NTA is present as a free residual. This is advantageous in that NTA can be determined more precisely than EDTA.

Suggested operating conditions for an 800-psi boiler under chelant treatment are listed in Table 6.4. Once again it is important to emphasize that a chelant treatment must be much more carefully controlled than those based on phosphate. If properly applied to suitable feed water, however, a chelant program can achieve cleaner boilers with less carryover of boiler salines and less disturbance by oil contamination than is normally realized with phosphate treatments.

At pressures above 900 psi caustic gouging becomes a serious consideration in the chemical treatment of boiler water; above 1200 psi it is a defi-

TABLE 6.4

Chelant Treatment in 800-psi Boilers

Component in boiler water	Suitable concentrations
Free Alkalinity, $(2P - M)$, (ppm $CaCO_3$)	75–150
Total Chelant (ppm EDTA)	20 max
Total Chelant (ppm NTA)	30 max
Total Chelant with Phosphonate (ppm NTA)	8–10*
Polymer (ppm)	3–5
Sulfite (ppm SO_3^{--})	15–20
Hydrazine (ppm N_2H_4)	0.10–0.20
Silica (ppm SiO_2)	25 max

* Assuming less than 0.5 ppm $CaCO_3$ total hardness in feed water.

nite hazard. In the range of 850 to 1200 psi in boilers provided with a make up of superior quality (either demineralized or evaporated), caustic corrosion can be controlled by using the coordinated phosphate-pH treatment,[14] discussed in Section 5.4b. In this pressure range sodium sulfite tends to decompose to some extent, releasing sulfur dioxide, an acidic gas that can cause severe corrosion in the condensate system. At pressures higher than those now being considered, the decomposition becomes more rapid. Thus, in boiler water containing 20 ppm Na_2SO_3 there is ≈ 0.2 ppm of SO_2 in steam at 1550 psi and 0.5 ppm of SO_2 in steam at 2000 psi. As sulfurous acid is much stronger ($K_1 = 1.72 \times 10^{-2}$) than carbonic acid ($K_1 = 4.2 \times 10^{-7}$), the pH of condensed high-pressure steam could be < 2. Furthermore, the decomposition of sulfite increases the free alkalinity in the boiler water.

$$SO_3^{--} + H_2O = 2OH^- + SO_2 \qquad (6\text{-}7)$$

For these reasons, hydrazine is the oxygen scavenger of choice above 900 psi.

Because the coordinated phosphate-pH system is especially appropriate for marine boilers, which require a relatively small volume of make up prepared by the evaporation of seawater, this application will be used for illustration.

The distillation of seawater is accomplished by heating it with 10-psi steam, then introducing the hot saline water into one or more flash chambers where it vaporizes. The distillate is cooled and deaerated; the concentrated brine remaining is pumped overboard to the sea. Tubes in the feed water heater, the distillate cooler, and the shells of the flash chambers, all of which are exposed to seawater, are fabricated of 90/10 copper-nickel alloy. The distillate pump is usually made of cast iron with a phosphor-bronze impeller; the impeller of the brine pump is aluminum bronze. Leakage in the main condenser, which is cooled by passing seawater once-through aluminum brass tubes, and upsets in the operation of the evaporator, can introduce seawater into the boilers leading to scaling, tube failures, and damage to the turbine. When operating correctly, distilled water from a two-stage, flash-type evaporator contains 0.25-1 ppm of total dissolved solids with perhaps 0.05-0.07 ppm of silica.

In applying the coordinated phosphate-pH program a moderate residual of phosphate (15-25 ppm) is maintained in the boiler water, so that it is always available to react with calcium and magnesium contamination in the feed water. The pH of the boiler water is maintained between 9.8 and 10.2, which is high enough to ensure the precipitation of calcium and magnesium, but not so high that excess hydroxyl ion is present to cause caustic gouging, the most common type of corrosion in high-pressure boilers. The shaded area in Fig. 6.1 shows the ranges in which phosphate and pH should fall. This can be accomplished by adding disodium phosphate and/or sodium hydroxide. As can be seen by examining the arrows in the figure, adding disodium phosphate increases the phosphate residual without affecting the pH, whereas adding sodium hydroxide raises the pH, but does not affect the phosphate residual. The objective is to keep the pH and the concentration of phosphate within the normal operating region in Fig. 6.1, but in any case, always beneath the curve.

Seawater contains dissolved air, the greater part of which is transferred into the distillate in the evaporators. In addition, air can leak into portions of the condenser circuit that are under vacuum, if mechanical joints are not tight. As the main condenser on a ship is one of the larger pieces of equipment, and is of bolted construction, it is a primary source of air contamination. As in any steam plant, most of any dissolved gases are removed from the feed water by mechanical deaeration. Residual oxygen ($<$ 10 ppb if the deaerator is functioning properly) is reduced by injecting a solution of hydrazine between the high-pressure and low-pressure stages of the turbine when at sea, or into the storage section of the deaerator when in port. The turbine cross-over is the preferred point of injection when steaming, rather than the boiler feed water. If hydrazine were added to the feed, it

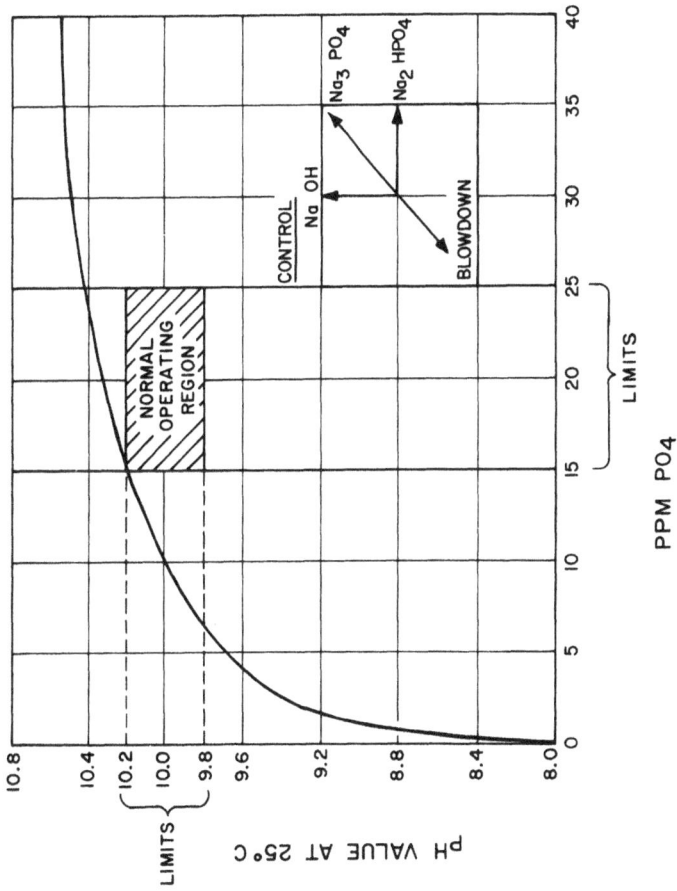

Fig. 6.1 Coordinated Phosphate-pH Curve

would subsequently be completely destroyed in the superheater. Thus, none would be present in the steam to reduce oxygen in the condenser, which being under vacuum, is susceptible to in-leakage of air. Then too, ammonia formed by the decomposition of residual hydrazine is present in the condenser. Injecting hydrazine into the turbine cross-over thus ensures that a protective residual is always present at the condenser inlet, uniformly distributed in the steam.

Corrosion by carbonic acid can be minimized by adding a neutralizing amine to the storage section of the deaerator. The amine (morpholine, cyclohexylamine, or a mixture of the two is satisfactory) vaporizes with steam in the boiler, passes through the turbine into the condenser circuit, and finally combines with carbon dioxide as the wet steam condenses. Despite this protective treatment some corrosion of steel, bronze, and brass inevitably occurs in the steam and condensate systems. Deck steam lines, which are used intermittently, are particularly susceptible to corrosion. When these lines lay idle rusting occurs, then when steam is turned into the lines loose rust is swept back through the atmospheric condenser into the drains tank.

Although the deaerator operates effectively at full steam load, it suffers a sharp drop in efficiency at partial load, or during changing loads as, for instance, when the ship is maneuvering. Significant concentrations of oxygen enter the boiler under these conditions, as it is not feasible to change the dosage of hydrazine continually. As a consequence, both steel and nonferrous surfaces are corroded to some extent, introducing finely divided oxides of iron and copper into the condensate system. These oxides, if allowed to enter the boiler, deposit in areas where the velocity of flow is low and at rough spots such as butt-welded tube joints, where they initiate pitting and gouging. For this reason, the filtration of condensate should be an integral part of water treatment on ships.

To remove the oxides of iron and copper, a cartridge filter designed to retain particles larger than 5 microns is installed in the condenser circuit. Ideally, the filter should be located in the main condensate stream just upstream of the deaerator, but as approximately 90 percent of the total oxides can be removed by filtering the contaminated drains, it is often sufficient to install the filter on the discharge side of the atmospheric drains pump. On many ships the flow in this system is significantly less than the total flow of condensate, so that a smaller, and therefore less costly, filter suffices. The filter should have the capacity to contain at least 100 lb of solid material to permit cartridge changes to be made at 6–9 month intervals. As a cartridge can be changed in about one-half hour, it is unnecessary to fit dual filters; the single filter can simply be bypassed while changing the cartridge. An AMF CUNO Type SL filter* is suitable for this service.

The following control limits for the coordinated phosphate-pH program are typical for marine boilers with operating pressures in the range of 850–1200 psi.

TABLE 6.5

Feed Water and Condensate for Coordinated Program

Component	Suitable concentrations
pH (Condensate)	8.6–9.0
Suspended Solids (Condensate)	No particulates larger than 5 microns
Hydrazine (ppm N_2H_4)	0.01–0.05*
Iron (ppm Fe)	0.02 max
Copper (ppm Cu)	0.02 max
Oxygen (Deaerator Outlet), (ppb O_2)	10 max
Ammonia (Condensate), (ppm NH_3)	0.5 max

* Measured in the feed water line after the last feed water heater. Hydrazine at this point should be at the lowest level detectable, but an actual residual should be measurable.

For reasons given in Chapter 5, page 87, the coordinated phosphate-pH method of caustic control is unsatisfactory at operating pressures exceeding 1200 psi. The ratio of sodium to phosphate should not exceed 2.6 in these high-pressure boilers if caustic gouging is to be avoided. The achievement of this ratio is called congruent control,[15] and is accomplished by maintaining a concentration of 2-6 ppm of phosphate with pH in the range of 8.5-9.3. Fig. 6.2 represents the relationship of pH to various concentrations of phosphate in a mixture containing 60 percent of Na_3PO_4 and 40 percent of Na_2HPO_4, in which the ratio of sodium to phosphate is 2.6. The normal operating region shown in Fig. 6.2 (2-4 ppm PO_4, pH 8.5-9.3) is suitable for boilers operating in the range of 1500–2000 psi; from 1200 to 1500 psi the pH range is the same, but the concentration of phosphate

* Available from AMF CUNO Division, Meriden, Connecticut 06450

should be 4-6 ppm. The proper ratio is obtained in an operating boiler by adding H_3PO_4, NaH_2PO_4, Na_2HPO_4, Na_3PO_4, or NaOH, as appropriate. In most operations small amounts of alkali carbonates unavoidably enter the boiler, so in some cases it is necessary to add phosphoric acid to bring the pH of the boiler water into the desired range. In the special case of leakage of sodium ion through the cation exchange section of a demineralizer, sodium dihydrogen phosphate should be used, rather than phosphoric acid, to neutralize the sodium hydroxide that appears in the effluent of the anion exchanger.

TABLE 6.6

Boiler Operation with Coordinated Program

Components in boiler water	Suitable concentrations
pH	9.8–10.2
P-Alkalinity (ppm $CaCO_3$)	12–22
M-Alkalinity (ppm $CaCO_3$)	25–45
Conductivity–Unneutralized (μmhos)	300 max[1]
Chloride (ppm Cl^-)	30 max[2]
Silica (ppm SiO_2)	4 max[3]
Phosphate (ppm PO_4^{---})	15–25
Hydrazine (ppm N_2H_4)	0.01–0.1

[1] If this value is exceeded, seawater contamination is indicated, and blowdown is required.

[2] Check for condenser leaks or evaporator upsets if this value is exceeded.

[3] Applies to marine boilers using evaporated make up. Higher values are allowable in stationary boilers (see Table 6.12), but on a ship would indicate a condenser leak or upset in the evaporator.

Virtually all steam generators that operate in excess of 1200 psi are used to produce electricity rather than process steam. The make up rate in these plants is, therefore, small—perhaps 1 or 2 percent—so raw water is purified either by evaporation or demineralization. In general, if the concentration of total dissolved solids in the raw water is < 500 ppm, demineralization is used; above 500 ppm, evaporation is likely to be more economical.

In boilers exceeding 800 psi, silica is nearly always the primary blow-

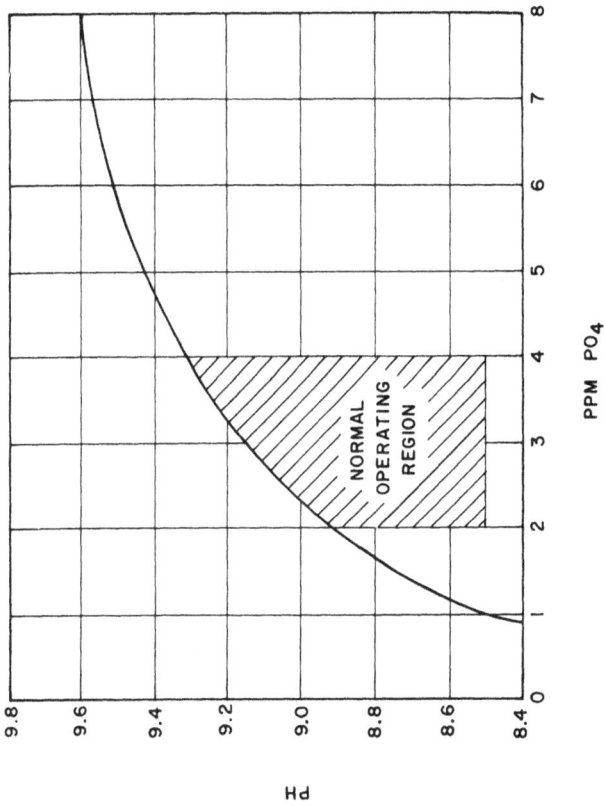

Fig. 6.2 Congruent control curve

down control so, in order to make economical use of water, it is necessary to use effective methods for reducing the concentration of silica in raw water. At 1200 psi the maximum concentration of silica that can be tolerated in boiler water is about 4 ppm SiO_2; at 2000 psi the limit is 1 ppm SiO_2, or a little less. As an illustration of the effect of silica, assume that the make up rate is 1 percent of the feed water to a 2000-psi boiler that must be operated at a blowdown ratio of 200. If the concentration of silica is to be limited to 1 ppm in the boiler water, its concentration in the feed water must be 1/200, or 0.005 ppm SiO_2. As condensate is normally silica-free, the concentration in the make up must be limited to 0.5 ppm SiO_2. This concentration is easily achieved by a simple demineralizer consisting of a strong-acid cation exchange resin in the hydrogen form, followed by a strong-base anion exchange resin in the hydroxide form; water can be demineralized this way at the rate of 2–5 gal/min/ft^3 of resin. As the operation of this conventional type of demineralizer is described in detail in Section 4.3b, it is only necessary to add here that if the M-alkalinity in the raw water to be demineralized is greater than 50 ppm $CaCO_3$, it is worthwhile to install a degasifier between the cation and the anion exchange units. This reduces the concentration of carbon dioxide in the effluent of the cation exchanger to about 10 ppm CO_2, thereby decreasing the load on the anion exchange resin and lowering the consumption of sodium hydroxide—now an expensive, energy-intensive chemical.

TABLE 6.7

Boiler Operation with Congruent Program

Components in boiler water	Suitable concentrations	
	1500 psi	2000 psi
pH	8.5–9.3	8.5–9.3
Phosphate (ppm PO_4)	4–6	2–4
Hydrazine (ppm N_2H_4)	0.05–0.10	0.05–0.10
Silica (ppm SiO_2)	3 max	1 max

At pressures exceeding 2500 psi, it is essential that no appreciable solids be introduced into the boiler. Accordingly, phosphates can no longer be used, and also a more efficient method of demineralization is demanded. Table

6.8 shows the performance of a typical sequence of ion-exchange resins used to process a flocculated river water having a conductivity of 200 μmhos; the flow rate in this particular train is 650 gal/min. The weak-base anion exchange resin used for the second stage of demineralization has a high exchange capacity and can be regenerated much more efficiently than a strong-base resin. The latter is required too, however, to remove silicate ion.

TABLE 6.8

Performance of a Four-Stage Demineralizer

Type of resin	Volume of resin (ft^3)	Effluent conductivity (μmhos)
Cation Exchange, Strong Acid	150	1200.00
Anion Exchange, Strong Base	70	3.00
Anion Exchange, Weak Base	80	1.00
Mixed Bed, Strong Acid-Strong Base	50	0.15

As only volatile chemicals can be used for water treatment above 2500 psi, hydrazine alone is added to the feed water at 150-200 percent of the concentration of oxygen, which should be less than 5 ppb. In some plants hydrazine solution is added at the discharge of the condensate extraction pump, and sometimes aqueous ammonia is injected into the suction of the boiler feed pump.

TABLE 6.9

Boiler Operation with Volatile Treatment

Components in boiler water	Suitable concentrations 2000-3000 psi
pH	8.5-9.5
Hydrazine (ppb N_2H_4)	20-30
Total dissolved solids (ppm)	2 max
Silica (ppm SiO_2)	0.5 max

At and above the critical pressure of 3203.6 psi, virtually no solids can be tolerated in boiler feed water, as all of this water is converted to steam that passes through the turbine. Copper, which enters the boiler as corrosion products, is the most troublesome contaminant in supercritical boilers. The metal and its oxides tend to form heavy deposits 10-20 mils thick on the wheels and stage diaphragms in high-pressure turbines. Pocock and Stewart[16] found mixtures of copper, cuprous oxide, and cupric oxide in the high-pressure stages of a 4500-psi turbine, with lesser amounts in the first and second reheat turbines and in the low-pressure turbines. These deposits do not form in superheaters, but because of the enormous flow rates in supercritical boilers, even a few parts per billion of copper oxides in the steam can soon drastically reduce the efficiency of a turbine. As these deposits cannot be washed off with water, they must be removed mechanically, e.g., by blasting with Alundum. Therefore, every effort is made to prevent metallic oxides from entering the boiler. This can be done by raising the pH of the condensate to prevent the corrosion of aluminum brass or admiralty brass in condensers, by filtration, and by demineralizing ("polishing") the condensate. By installing large mixed-bed ion exchangers, some with flow rates of 20,000 gpm or more, in the condensate system, part of the metallic oxides and colloidal hydroxides can be eliminated. The total concentration of iron can be reduced to 3-4 ppb, and that of copper to 1-2 ppb, with the conductivity of the effluent less than 0.1 μmhos. This process also protects the boiler and turbine in the event of a condenser leak. In demineralizing condensate account must be taken of the exhaustion of the cation exchange resin by ammonia or amines.

The mechanical retention in a demineralizer of iron oxides and other particulates is, of course, an incidental advantage, but the primary purpose of demineralizing feed water is to prevent contamination of the turbine. In most condensate systems it is advantageous, if not essential, to install a leaf filter combined with a cellulose filter-aid downstream of the condenser hotwell, and upstream of the condensate demineralizer. Condensers are usually fabricated of copper alloys rather than steel on account of the poor resistance of the latter to aerated cooling water. Stainless steel can be used if the coolant is fresh water, but is inapplicable in locations where brackish water or seawater is used, because of the high susceptibility of Austenitic alloys to stress corrosion cracking by chloride ion. Pocock, et al.[17] recommend that feed water heaters be fabricated with carbon steel to avoid deposits of copper oxides in turbines. Ammonia is then added to the feed water to raise its pH to 9.4, at which value the corrosion rate of iron is at a minimum. These authors also found that a large proportion of the corrosion products of iron is smaller than 0.45 micron. As particles this small are not retained on ordinary filters, it is important to eliminate the corrosion of iron by carbon dioxide in the condensate system by elevating the pH of the water. Generally, concentrations of iron < 5 ppb cannot be removed by demineralization or conventional filtration.

Duff and Levendusky[18] have described the purification of condensate by the Powdex* process, in which a Nylon-wound filter cartridge is coated with a mixed ion-exchange resin, thus providing both efficient filtration and demineralization in a single unit. When a breakthrough of silica occurs, the exhausted resin is discarded and the filter cartridges are recoated with fresh resin. This purification train, installed just ahead of the deaerator, eliminates deposits of silica in the turbines, reduces the concentration of metallic oxides in boiler feed water, and removes soluble salts introduced by condenser leaks. Grant and Crouse[19] report that Powdex reduced the concentrations of iron and copper by 48 percent and 70 percent, respectively, and eliminated 30–40 percent of colloidal silica in boiler feed water. Care must be taken, however, to minimize the leakage of resin into boilers, as it lowers the pH of the boiler water and increases the concentration of ammonia in the steam.

TABLE 6.10

Operating Conditions for Supercritical Boilers

Components in boiler water	Suitable concentrations
pH	9.3–9.5
Hydrazine–feed water (ppb N_2H_4)	10–15
Ammonia (ppb NH_3)	250–400
Silica (ppb SiO_2)	20 max
Sodium (ppb Na)	1 max
Total dissolved solids (ppb)	20 max
Iron–steam (ppb Fe)	10 max
Copper–steam (ppb Cu)	5 max

* Trademark of the Graver Water Division, Ecodyne Corporation, U.S. Highway 22, Union, New Jersey 07083.

TABLE 6.11

Recommended Feed Water Quality

Pressure (psi)	Silica range (ppm)	Maximum (ppm)			
		Total hardness	Oxygen	Iron*	Copper*
100	15-25	75.00	—	—	—
200	10-20	20.00	—	—	—
300	7.5-15	2.00	—	—	—
500	2.5-5.0	2.00	0.030	—	—
600	1.3-2.5	0.20	0.030	—	—
750	1.3-2.5	0.10	0.030	0.050	0.020
900	0.8-1.5	0.05	0.007	0.020	0.015
1000	0.2-0.3	0.05	0.007	0.020	0.015
1500	0.3 max	0.00	0.005	0.010	0.010
2000	0.1 max	0.00	0.005	0.010	0.010
2500	0.05 max	0.00	0.003	0.003	0.002
3200+	0.02 max	0.00	0.002	0.002	0.001

* In modern industrial boilers, which have extremely high rates of heat transfer, these concentrations should be essentially zero. Similarly, total hardness should not exceed 0.3 ppm $CaCO_3$, even at the lower pressures; suspended solids in the feed water should be zero, if possible.

TABLE 6.12

Recommended Concentrations of Boiler Salines

Pressure (psi)	Saturation temperature (F)	Maximum (ppm)						Range (ppm)		
		Dissolved solids	Suspended solids*	Total alkalinity**	Silica	Sludge conditioners Natural	Synthetic	Residual phosphate	Residual sulfite	Residual hydrazine
100	328	5000.00	500	900	250.00	150	15	NR†	90–100	NR
200	382	4000.00	350	800	200.00	150	15	40–50	80–90	NR
300	417	3500.00	300	700	175.00	100	15	30–40	60–70	NR
500	467	3000.00	60	600	40.00	70	15	25–30	45–60	NR
600	486	2500.00	50	500	35.00	70	10	20–25	30–45	NR
750	510	2000.00	40	300	30.00	NR	10	15–20	25–30	NR
900	532	1000.00	20	200	20.00	NR	5	10–15	15–20	0.10–0.15
1000	545	500.00	10	50	10.00	NR	3	5–10	NR	0.10–0.15
1500	596	150.00	3	0	3.00	NR	NR	3–6	NR	0.05–0.10
2000	636	50.00	1	0	1.00	NR	NR	1–3	NR	0.05–0.10
2500	668	10.00	0	0	0.50	NR	NR	NR	NR	0.02–0.03
3200+	705	0.02	0	0	0.02	NR	NR	NR	NR	0.01–0.02

* Guidelines for pressures from 100 to 900 psi apply to conventional field-erected boilers with moderate rates of heat transfer, say, 50,000 Btu/h/ft². At high rates characteristic of packaged boilers, large amounts of insoluble material cannot be managed effectively by an dispersant presently available.
** Zero alkalinity refers to hydroxide ion, i.e., (2P − M). There is always some alkalinity produced by ammonia, hydrazine, morpholine, or other bases.
† NR = Not recommended.

For convenience, the many values quoted in the foregoing discussion of water treatments appropriate for boilers at specific operating pressures are summarized in Table 6.11 and Table 6.12. These tables were prepared as guidelines only. Values shown reflect the assumption that boilers operated at low pressure can use poorer feed water than those operated at higher pressures. Thus, feed water softened by the hot-lime process contains residual hardness of 15-20 ppm $CaCO_3$. This would be satisfactory for an older boiler generating process steam at 200 psi, but would be unusable at the same pressure for a modern packaged boiler, in which high concentrations of suspended solids are intolerable. Similarly, modern marine boilers at 850-1200 psi have very little tolerance for suspended solids; for these applications the values in Table 6.12 are too high.

a. The Water-Treating Plant

It is apparent from the preceding discussion that there isn't much in the way of chemical treatment of feed water for extremely high-pressure boilers: their make up is evaporated or demineralized raw water to which hydrazine is added. With few exceptions, boilers operated at more than 1200 psi are used in central power stations for generating electricity. Industrial steam plants, on the other hand, generate process steam in boilers at moderate pressures that do not require exceptionally pure feed water. Because the operation of steam plants is a matter of considerable importance in the manufacturing industries, it is important to consider a typical water-treating plant suitable for preparing make up for 850-psi boilers. Proper operating conditions for such boilers, using the caustic reserve method of treatment (see pp. 125-126) are shown in Table 6.13.

TABLE 6.13

Operating Conditions for 850-psi Boilers

Components in boiler water	Suitable concentrations
Total dissolved solids (ppm)	1500 max
Silica (ppm SiO_2)	25 max
M-Alkalinity (ppm $CaCO_3$)	250 max
Phosphate (ppm PO_4)	10–15
Sulfite (ppm SO_3)	15–20

It is proposed to treat raw water having the composition shown in Table 6.14, to make it suitable for make up for six 850-psi boilers, each producing 175,000 lb of steam per hour.

TABLE 6.14

Analysis of Raw Water

Component	Abbreviation	Concentration
Total hardness (ppm $CaCO_3$)	TH	79.0
Calcium hardness (ppm $CaCO_3$)	CaH	47.0
Magnesium hardness (ppm $CaCO_3$)	MgH	32.0
P-Alkalinity (ppm $CaCO_3$)	P	4.0
M-Alkalinity (ppm $CaCO_3$)	M	65.0
Sodium (ppm Na)	Na	15.0
Chloride (ppm Cl)	Cl	7.0
Sulfate (ppm SO_4)	SO_4	17.0
Silica (ppm SiO_2)	SiO_2	12.0
Total dissolved solids (ppm)	TDS	160.0
Conductivity (μmhos)	Cond	265.0
pH	pH	8.4

In order to conserve water and fuel, it is necessary to operate the boilers at not more than 5 percent blowdown. To achieve this rate it is necessary to reduce the concentrations of silica and total dissolved solids, both of which can be accomplished by cold-lime softening in a Spaulding precipitator. As explained in Chapter 4, pp. 57-58, it is best when using a Spaulding precipitator, to arrange the chemical treatment for maximum removal of silica, then complete the softening by cation exchange.

Fig. 6.3 shows the components, chemicals added, and sequence of operations in a water-treating plant designed for preparing make up for boilers operated at moderate pressure. Raw water of the composition shown in Table 6.14 flows into the inner cone of a Spaulding precipitator and mixes with the proper dosages of the flocculating chemicals listed in Fig. 6.3. Effluent water, at a pH of about 10.2, is next filtered through graded anthracite coal to remove any particles of floc carried over from the precipitator. The pH of the water is lowered to 8.3 by injecting sulfuric acid, then the water flows through cation exchange resin in the sodium form, which re-

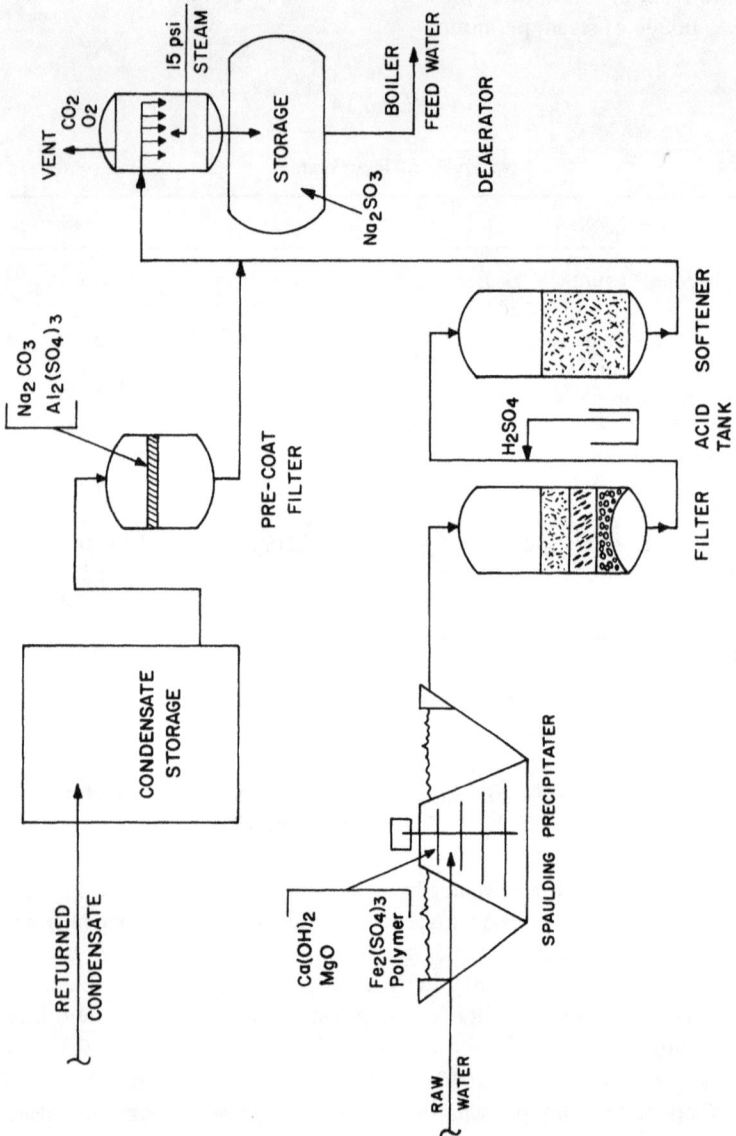

Fig. 6.3   Water treatment plant for moderate pressure boiler feed water.

places calcium and magnesium ions with sodium ions.

Returned condensate is passed through a filter precoated with aluminum hydroxide gel to remove any oil introduced by machinery or process leaks. Filtered condensate is then combined with the filtered, softened raw water and deaerated. Catalyzed sodium sulfite is added to the storage section of the deaerator to scavenge the last traces of oxygen, after which the combined feed water is ready for use. Chapter 3 and 4 contain complete explanations of all these operations, but it is worth while here to examine the operation of the Spaulding precipitator more closely, as these units tend to be troublesome if not properly managed. Fluctuations in the composition of the raw water also upsets these units.

In order to calculate the proper dosage of treating chemicals it is necessary to have a way of measuring the flow of water through the precipitator. A flow meter is usually provided in the raw water feed line that includes both an instantaneous reading recorded on a 24-h circular chart and a totalizer. Charts normally are marked from 0 to 100; an integrating factor is required to convert the chart reading to gallons per minute. Thus, if the integrating factor is, e.g., 15, and the chart reading happens to be 52, then the instantaneous rate of flow is 15 × 52, or 780 gpm. If, for example, the totalizer at a certain time reads 2,748,403, and 8 h later reads 2,748,915, the number of gallons processed in 8 h is the difference in readings, 512, multiplied by 900, i.e., 15 × 60, or 460,800 gal.

In order to reduce the concentration of silica, the proper amount of slaked lime, $Ca(OH)_2$, is added for lime softening together with magnesium oxide at three times the concentration of silica in the raw water. Referring to Table 6.12 it is seen that M > CaH, therefore the proper dosage of $Ca(OH)_2$ is 0.742(M + MgH), or 0.742(65 + 32) = 72 ppm $Ca(OH)_2$ [see Chapter 4, pp. 41-43]. The concentration of silica in the raw water is 12 ppm, so 3 × 12, or 36 ppm of MgO should also be added. These two chemicals can be mixed together and fed continuously as a slurry (approximately 5 percent lime) to the inner cone, proportioned to the flow of water through the precipitator. To contain the mixed slurry a cylindrical mixer 36 in. deep with a radius of 4 ft, equipped with an agitator to keep the slurry uniformly suspended is used. The method used to prepare this mix and how much of it should be fed in a given length of time, e.g., 8 h, to achieve the desired concentrations of $Ca(OH)_2$ and MgO must now be determined. Also, it is convenient to use entire bags rather than odd numbers of pounds in making the mix; lime is available in 100-lb bags, magnesium oxide in 60-lb bags.

First, the volume of the mixer is $h\pi r^2$, or $3 \times \pi \times 16 = 150.8$ ft^3, equivalent to 1130 gal. If 500 lb of lime is added and diluted to 36 in. with water, the resulting slurry contains about 5 percent of lime. Assuming an average flow of 460,800 gal, equivalent to 3,840,000 lb of water in 8 h, it is necessary to add $72 \times 3,840,000/10^6$, or 277 lb of Ca(OH)$_2$ continuously over an 8-h period. This can be done by adjusting the proportioner to feed $277 \times 36/500$, or 20 in. of slurry over an 8-h period. Similarly, to obtain 36 ppm of MgO it is necessary to add $36 \times 3,840,000/10^6$, or 138 lb of MgO in 8 h. As this is to be fed in the same 20 in. of slurry, $138 \times 36/20$, or 248 lb of MgO must be added to the lime mixer also. Note that four bags of MgO weighs a total of 240 lb, which is close enough for this operation. Next the alkalinity differential, (2P − M), of the effluent from the precipitator must be adjusted so that it falls in the range of 12-18 ppm CaCO$_3$. This is done with a solution of ferric sulfate fed from a separate mixer with its own proportioner; 200 lb of ferric sulfate in 36 in. of water is satisfactory. The usual dosage is around 7 ppm of Fe$_2$(SO$_4$)$_3$ − about 5 in. of solution in 8 h, if using a 36-in. mixer.

The heat of solution of anhydrous ferric sulfate is such that about 50,000 Btu's are liberated when a 200-lb portion of the salt is dissolved in water. The corresponding rise in temperature of the water increases the rate of solution of the salt, but the final temperature should not exceed 100 F, or fuming occurs. If the temperature of the water is too low, or if the proportion of water to salt is too great, undissolved ferric sulfate is likely to foul the mixer and feeding equipment. It is, therefore, important to take into account the temperature of the dilution water and its volume when preparing an aqueous solution of ferric sulfate. The following table lists volumes of water at various temperatures to use for dissolving 200 lb of ferric sulfate; at each temperature listed solution is complete in $\approx$ 20 min, with a final solution temperature of 95-100 F. After the salt is completely dissolved, dilute to the working volume of the mixer (36 in. in this example).

Once the proportioners on the lime and ferric sulfate mixers are properly set for a particular flow of water, they automatically add the proper amounts of the slurries as the water flow changes because of more or less steam demand, or fluctuations in the amount of returned condensate.

Finally, a nonionic polymer emulsion* is used to control carryover in the precipitator; 2 ppm is the usual dosage, and it must be added continu-

---

* Magnifloc 1906N, available from American Cyanamid Company, Industrial Chemicals and Plastics Division, Wayne, N.J. 07470, is suitable.

TABLE 6.15

Preparation of Ferric Sulfate Solutions

Temperature of water (F)	Volume of water (gal)	Percent salt (wt/wt)
45	70	28
60	95	20
70	120	17
80	240	9
90	360	6

ously from its own mixer to the inner cone of the precipitator. Approximately 3 gal of the emulsion can be mixed with water in a 55-gal drum and fed into the inner cone by means of a chemical feed pump adjusted to empty the drum in 24 h. The rate of addition of the polymer and the rate of desludging the precipitator are adjusted so that the surface of the sludge blanket is 6–7 ft below the surface of the water flowing into the collection trough. Sodium aluminate is often added to the lime slurry as a coagulant in lime softening, but the polymer recommended here is more efficient and convenient to use when the aim is to remove silica. If filter alum is used it must be added from a separate mixer rather than in the lime slurry.

We recall now that the water-treating plant under discussion is to provide make up for six 850-psi boilers, each generating 175,000 lb of steam per hour. This amounts to $175,000 \times 6 \times 24$, or 25,200,000 lb of steam per day, which is equivalent to about 3,000,000 gal of water. Because of the advantages in operating these boilers at 5 percent blowdown, the blowdown ratio will be $1/0.05 = 20$. From Eq. (1-6) the rate of blowdown is 3,000,000/19, or about 158,000 gal/day, and from Eq. (1-1), the total feed water flow is (3,000,000 + 158,000), or 3,158,000 gal/day. This volume of water passes through the deaerator every 24 h where it must be treated with 1–2 ppm of sulfite to reduce any residual oxygen in the feed water. Catalyzed sodium sulfite is proportioned from a mix tank in the same way as lime and ferric sulfate. To add 1 ppm of SO_3^{--} to the water, 1.6 ppm of Na_2SO_3, i.e., $1.6 \times 3,158,000 \times 8.34/10^6 = 42$ lb of Na_2SO_3 every 24 h. This assumes, of course, that the deaerator is working properly and producing water containing less than 10 ppb of dissolved oxygen.

The removal of oil from condensate using precoated filters is described in sufficient detail in Section 3.3 so that it need not be considered here, except to note that these filters also help to remove particulate corrosion products, especially if cellulose fiber is included as a filter aid. Diatomaceous earth is not satisfactory for this purpose, as silica is leached from it by hot condensate, subsequently contaminating the feed water.

As already noted, corrosion and corrosion products in steam and condensate systems lead to a number of undesirable consequences including damage to feed water heaters, pumps, and boilers, as well as plugging steam traps and perforating piping. This corrosion is caused by carbon dioxide, which is a decomposition product of carbonates in the boiler, and of oxygen, which is introduced in air drawn into the condensate system through steam vents, condensate extraction pumps, and bolted condensers, or in undeaerated quench water. These gases are not corrosive when dispersed in dry steam, but as condensation commences, droplets of water form (particularly at bends where steam lines change direction sharply) and corrosion proceeds. Brindisi[20] reports that corrosion by carbon dioxide alone is mild at concentrations less than 6 ppm of CO_2 when the pH is $\approx$ 7. Attack is aggressive, however, below pH 5.8 at concentrations exceeding 20 ppm of CO_2; if oxygen is also present the concerted attack is much more intense than that by either alone. Carbon dioxide corrosion is indicated by a characteristic etching and grooving in the metal surface, whereas the ravages of oxygen are pitting and tuberculation.

Filming amines are often recommended for protection against oxygen. Added as an emulsion to the feed water, they are said to be effective at pressures up to 2200 psi and superheater temperatures to 1005 F. The best filming is obtained in systems in which condensate contains < 10 ppm of total solids, with pH in the range of 4-8. Slow flow rates, pH values above 9, temperature less than 120 F, corrosion products, and hydrocarbon contamination in condensate all interfere with the formation of films. For these reasons, filming amines are inappropriate in many industrial steam plants. As this method of treatment is widely used, however, it is worthwhile to mention some important precautions that should be observed in applying these chemicals. Emulsified filming amines are injected continuously into the feed water downstream of the deaerator, or alternatively, into the steam header. The pH of the water used to dilute the emulsion must be < 9. Treatment is started at 0.5 ppm, then increased by 0.5 ppm every two weeks until a maximum dosage (based on steaming rate) of 2 ppm is reached. If the condensate becomes heavily contaminated with corrosion debris, this water should be discarded until the system is cleared.

Neutralizing amines, in general, are easier to apply and have fewer disadvantages than filming amines. The dosage used depends upon a number of items, including the degree of corrosion protection required, the percentage of returned condensate, the equivalent weight of the amine formulation, and the pH to be maintained in the feed water. In the utilities industry a pH of 8.8–9.2 has been fairly well established as adequate to protect copper and nickel alloys, although Pocock, et al.[21] found that the optimum protection of iron was achieved at a pH of 9.3–9.4. In contrast to filming amines, neutralizing amines do not dislodge massive amounts of old corrosion products in a dirty system, so treatment can be started at any dosage. Methods for estimating the proper dosage are given in Chapter 5, pages 108–109.

Before proceeding to a discussion of the internal chemical treatment of the 850-psi boilers being used as an example, it is of interest to examine the effect of the addition of morpholine on the concentrations of iron and copper in condensate. Bonafede and Sandell,[22] in a study of corrosion in a feed water system containing cupro-nickel high-pressure heaters, found a preponderance of suspended over dissolved iron. They also observed that, in the presence of morpholine, the concentrations of both suspended and dissolved iron were reduced, whereas the concentration of dissolved copper was increased. Suspended copper was not significantly affected. Table 6.16 summarizes a few of these results.

TABLE 6.16

Effect of Morpholine on Iron and Copper in Condensate

Condensate source		Iron (ppb)			Copper (ppb)		
		Susp.	Diss.	Total	Susp.	Diss.	Total
Turbine	A	7.7	1.9	9.6	1.2	0.9	2.1
	B	3.4	1.5	4.9	1.0	3.2	4.2
Process No. 1	A	27.0	2.5	29.0	0.2	0.9	1.1
	B	4.2	1.7	5.9	0.2	1.0	1.2
Process No. 2	A	25.0	3.6	29.0	0.3	0.9	1.2
	B	4.5	1.7	6.2	0.2	1.1	1.3

A = Without morpholine
B = With morpholine

b. Internal Treatment

Having completed a description of the external treatment of make up water for 850-psi boilers, their internal treatment can now be discussed. Before proceeding with this, however, the answers to four questions must be determined.

1. What are the percentages of make up and condensate in the feed water?
2. What is the minimum rate of blowdown achievable consistent with Table 6.13?
3. At this minimum blowdown, what is the rate of depletion of a chemical, e.g., phosphate, added to the boiler?
4. At the minimum blowdown, what is the rate of concentration in the boiler of salts in the feed water?

To answer the first two questions the chemical analyses of treated make up, condensate, and feed water must be referred to; suppose these are as shown in Table 6.17.

TABLE 6.17

Analyses of Water Streams

Component	Make up	Condensate	Feed water
P-Alkalinity (ppm $CaCO_3$)	0.0	1.0	1.0
M-Alkalinity (ppm $CaCO_3$)	6.0	2.0	4.0
Silica (ppm SiO_2)	1.8	0.2	0.9
Total dissolved solids (ppm)	73.0	5.0	35.0

The volume of treated make up plus the volume of condensate together comprise the total boiler feed water. By accurately measuring the concentration of silica in each of these streams, it is possible to calculate the percentage of make up and of condensate in the feed water. The volumes of make up, condensate, and feed water are designated as V_m, V_c, and V_f, respectively. Table 6.17 indicates that the concentration of silica in the make up is 1.8 ppm, in the condensate it is 0.2 ppm, and in the feed water it is 0.9 ppm. Therefore,

$$0.9V_f = 1.8V_m + 0.2V_c$$

Also,

$$V_f = V_m + V_c$$

V_c is eliminated by multiplying the second equation by -0.2, then adding the two equations.

$$
\begin{aligned}
0.9V_f &= 1.8V_m + 0.2V_c \\
-0.2V_f &= -0.2V_m - 0.2V_c \\
\hline
0.7V_f &= 1.6V_m
\end{aligned}
$$

Or

$$V_m/V_f = 0.7/1.6$$

From which

$$\text{Percent } V_m = 0.7 \times 100/1.6 \cong 44 \text{ percent}$$
$$\text{Percent } V_c = 100 - 44 \cong 56 \text{ percent}$$

Alternatively, the percentages can be calculated by letting m = decimal fraction of make up and c = decimal fraction of condensate in the feed water. Also, $m + c = 1$. Then,

$$1.8m + 0.2c = 0.9$$
$$1.8(1 - c) + 0.2c = 0.9$$
$$c = 0.56$$
$$m = 0.44$$

The two methods of calculation are, of course, equivalent. To answer the second question the values in Table 6.18 are compiled.

TABLE 6.18

Blowdown Limitations in 850-psi Boilers

Limits in Table 6.13		Feed water	Blowdown ratio	Percent blowdown
Total dissolved solids (ppm)	1500 max	35.0	43	2.3
M-Alkalinity (ppm CaCO$_3$)	250 max	4.0	62	1.6
Silica (ppm SiO$_2$)	25 max	0.9	28	3.6

To restrict the concentration of silica to a maximum of 25 ppm in the boilers requires a minimum blowdown of 3.6 percent. At this rate total dissolved solids is ≈ 970 ppm and M-alkalinity is ≈ 110 ppm CaCO$_3$ in the boiler water.

Before addressing the third question it is necessary to correct the preliminary values of f and b obtained on page 149 by assuming a blowdown rate of 5 percent. Presuming that the steaming rate, s, remains constant at 175,000 lb/h in each of the six boilers, b is calculated from Eq. (1-6).

$$s = b(R - 1) \qquad (1\text{-}6)$$
$$3,000,000 = b(28 - 1)$$
$$b = 111,000 \text{ gpd}$$

It is now possible to calculate the rate of depletion of phosphate injected into the steam drums of the six boilers. To do this Eq. (1-10) is used:

$$\log_e(c/c_o) = -b(t - t_o)/V \qquad (1\text{-}10)$$

Setting aside for a moment the mechanics of injecting phosphate, it can be simply assumed that each of the six boilers contains initially 15 ppm of phosphate, which will be called c_o. As b is the total blowdown, V must be the combined volume of the six boilers, i.e., 25,000 $\times$ 6 = 150,000 gal (see page 9). Time, t, is in days, therefore, if no phosphate is added for 24 h its concentration, c, at the end of one day will be

$$\log_e c - \log_e 15 = -111,000/150,000$$
$$\log_e c = 2.708 - 0.740$$

$$= 1.968$$

$$c = 7.1 \text{ ppm PO}_4$$

Thus, to keep the concentration of phosphate at 15 ppm it is necessary to add the equivalent of 7.9 ppm of phosphate every 24 h. In pounds of Na_2HPO_4 this is

$$7.9 \times 150{,}000 \times 8.34/10^6 \times 0.669 \cong 15 \text{ lb}$$

(Na_2HPO_4 contains 66.9 percent of phosphate.)

In practice a single solution of phosphate is prepared in a small mixer. The solution is injected directly into the steam drum of each boiler by an individual chemical pump having an adjustable stroke. Because of the pronounced tendency of phosphate to plug lines, it is desirable that the pumps run continuously, rather than being actuated by a percentage timer. Also, it is convenient to add sludge dispersants and antifoams, neither of which can be determined easily by chemical analysis, in the phosphate solution. For example, if the maintenance of 5 ppm of sodium polyacrylate and 3 ppm of antifoam in the boiler water is desired, it can easily be calculated, again using Eq. (1-10), that 2.6 ppm of polyacrylate and 1.6 ppm of antifoam must be added every 24 h. These concentrations are equivalent to 3.2 lb and 2.0 lb, respectively.

In contrast to steam generators in central power stations, the demand on industrial steam plants tends to be highly variable. As carryover is increased by a rapid rise in steam load, with the resulting fluctuations in water levels, it is advisable to use an antifoam in process steam plants. Packaged boilers, which have extremely high heat flux, and any boiler carrying high free alkalinity (B-reading > 8), should always be treated with antifoam. Two types are available: the soluble polyoxyalkylene glycols and the insoluble polyamides. The latter are seldom inactivated by normal boiler sludge, but they do tend to form sticky deposits in mixers, pumps, and transfer lines. Neither type is effective in the presence of organic contamination, but under normal operating conditions the polyoxyalkylene glycols are more efficient and convenient to use. An excellent antifoam can be prepared by diluting 5 gal of UCON 50 HB 5100* to 50 gal with water, then adding the dilute solution to the phosphate mixer in an amount sufficient to produce 3 ppm of the dilute solution ($\approx$ 0.3 ppm active) in

* Manufactured by Union Carbide Corporation, Chemicals and Plastics Division, 100 Oceangate, Long Beach, California 90802.

the boiler water.

In addition to an antifoam, it is advisable to include a protective colloid when using a phosphate treatment; this too can be included in the phosphate mix. Natural dispersants include quebracho, tannins, lignins, and mannuronic acid. The latter is obtained as a byproduct in the process of extracting iodine from seaweed. The major component (10–25 percent, dry weight) of the cell walls of brown algae (kelp) is a polyuronic acid, variously called algin, norgine, and alginic acid. Mannuronic acid is the polymerizing unit of alginic acid, with about 100 units per molecule. Synthetic polymers such as sodium carboxymethylcellulose or sodium polyacrylate are also used.

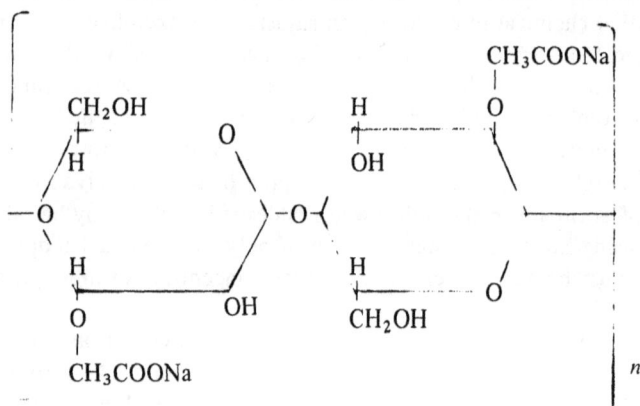

Sodium carboxymethylcellulose

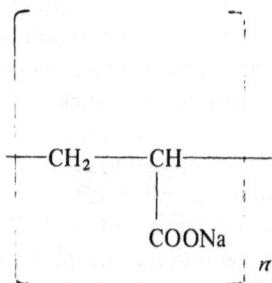

Sodium polyacrylate

The polyacrylates are more stable than carboxymethylcellulose at high temperature and alkalinity because the polymer is held together by carbon-carbon rather than carbon-oxygen bonds. Also, polyacrylates function at low concentrations—about 5 ppm in an 850-psi boiler.

It is now of interest to investigate what happens to the concentration of a salt in the boiler water, say silicate, that is also present in the feed water at some concentration, c_f. The concentration of the salt in the boiler water can increase or decrease depending upon the rate of blowdown. If a material balance on the boiler is made it is found that the change in total weight, W, of silica (or any other soluble salt in the feed water) in one hour is

$$W = Vc - bc + fc_f \qquad (6\text{-}8)$$

where

V = total volume of water in boiler (gal)
c = concentration of silica in boiler water (lb/gal)
c_f = concentration of silica in feed water (lb/gal)
b = blowdown rate (gal/h)
f = feed rate (gal/h)

The rate of change of concentration of silica is described by the following differential equation, which reflects the physical facts that the concentration is increased by entering feed water and decreased by blowdown.

$$dc/dt = fc_f/V - bc/V \qquad (6\text{-}9)$$

Assuming that the term fc_f/V is constant, which under any particular set of operating conditions it is, the variables are first separated:

$$dc = [fc_f/V - bc/V] dt$$

or

$$\frac{dc}{[fc_f/V - bc/V]} = dt \qquad (6\text{-}10)$$

Noting the form of the expression on the left, it is found in a table of integrals that

$$\int dx/(a + bx) = (1/b)\log_e(a + bx)$$

and thus,

$$\int_{c_o}^{c} \frac{dc}{[fc_f/V - bc/V]} = \frac{-1}{b/V}\log_e[fc_f/V - bc/V]\Big]_{c_o}^{c} \tag{6-11}$$

Hence,

$$\log_e \frac{[fc_f - bc]}{[fc_f - bc_o]} = -b(t - t_o)/V \tag{6-12}$$

or

$$fc_f - bc = (fc_f - bc_o)e^{-b(t-t_o)/V} \tag{6-13}$$

Eq. (6-13) for some purposes is more convenient to use if solved for c.

$$c = fc_f/b + (c_o - fc_f/b)e^{-b(t-t_o)/V} \tag{6-14}$$

As t becomes large the exponential term becomes vanishingly small and Eq. (6-14) reduces to Eq. (1-3).

$$c/c_f = f/b = R \tag{1-3}$$

As an example of the use of Eq. (6-12), suppose it is necessary to calculate the length of time it will take for the concentration of silica to reach 25 ppm in one of the 850-psi boilers, assuming it is just being started up, with the concentration of silica in the boiler water equal to that in the feed water, i.e., 0.9 ppm. The values to be substituted are:

$$f = 3,111,000/24 \times 6 = 21,600 \text{ gph}$$
$$b = 111,000/24 \times 6 = 771 \text{ gph}$$
$$V = 25,000 \text{ gal}$$
$$c = 25 \text{ ppm SiO}_2$$
$$c_o = c_f = 0.9 \text{ ppm SiO}_2$$

Therefore,

$$\log_e \frac{[21,600 \times 0.9 - 771 \times 25]}{[21,600 \times 0.9 - 771 \times 0.9]} = -771t/25,000$$

$$\log_e(0.0088) = -0.0308t$$

$$-4.73 = -0.0308t$$

$$t \cong 153 \text{ h}$$

If for some reason, one wants to know what the concentration of silica would be in 48 h, Eq. (6-14) would be used:

$$c_{48} = (21{,}600 \times 0.9/771) + (0.9 - 21{,}600 \times 0.9/771)e^{-771 \times 48/25{,}000}$$

$$= 25.2 + (0.9 - 25.2)e^{-1.48}$$

$$= 25.2 - (24.3 \times 0.227)$$

$$= 19.7 \text{ ppm}$$

As another example of the use of Eq. (6-12) suppose, through inattention, the concentration of silica in one of the 850-psi boilers rises to 32 ppm SiO_2. How long will it take to return the concentration to 25 ppm SiO_2, if the blowdown rate is increased to 1000 gph? (Note that when b is increased f increases by the same amount, i.e., $f = 21{,}829$ gph.)

$$\log_e \frac{[21{,}829 \times 0.9 - 1000 \times 25]}{[21{,}829 \times 0.9 - 1000 \times 32]} = -1000t/25{,}000$$

$$\log_e(0.433) = -0.040t$$

$$-0.837 = -0.040t$$

$$t = 20.9 \text{ h}$$

The foregoing examples show that the application of elementary mathematical analysis to the operation of a boiler yields a variety of useful quantitative information; a number of similar study problems are included in an appendix.

In Section 5.5 the volatility of silica was discussed and it was noted that the extent of volatilization depends upon the pressure, steaming rate, alkalinity, and the concentration of silica in the boiler water. Barker[23] s published a curve showing the relationship between the concentration of silica in boiler water and the pressure of the boiler necessary to keep the concentration of silica in steam at or below the maximum acceptable value of 0.02 ppm. Using the methods of analytic geometry, the following empirical equation can be derived from the curve.

$$P = -774 \log_{10}(\text{ppm SiO}_2) + 1948 \qquad (6\text{-}15)$$

Thus at 1500 psi the ppm of SiO_2 that can be carried in the boiler water is calculated as follows:

$$\log_{10}(\text{ppm SiO}_2) = (1500 - 1948)/(-774)$$

$$= 0.58$$

$$\text{ppm SiO}_2 = 3.8$$

Coulter, et al.[24] have presented data relating the ratio of silica in steam to that in boiler water to the alkalinity of the boiler water. The following empirical equation fits their data reasonably well.

$$\log_{10}R = 0.0106t \, F - 0.17(\text{epm OH}^-)^{\frac{1}{2}} - 8.27 \qquad (6\text{-}16)$$

where:

$$R = \text{ppm SiO}_2 \text{ in steam/ppm SiO}_2 \text{ in boiler water}$$
$$t \, F = \text{saturation temperature of steam}$$
$$\text{epm OH}^- = \text{equivalents per million of hydroxyl ion}$$

Suppose: $P = 850$ psi; $t \, F = 525$; ppm SiO_2 (in boiler water) $= 25$; P-alkalinity $= 140$; M-alkalinity $= 201$; pH $= 11.2$. Epm OH^- can be calculated from pH or from $(2P - M)$ as follows:

$$\text{pH} = 11.2$$

$$\text{pOH} = 14.0 - 11.2 = 2.8$$

$$(\text{OH}^-) = 10^{-2.8} = 1.58 \times 10^{-3} \text{ epl}$$

$$1.58 \times 10^{-3} \times 10^3 = 1.58 \text{ epm OH}^-$$

or,

$$(2P - M) \times 0.02 = \text{epm OH}^-$$

$$(2 \times 140 - 201) \times 0.02 = 1.58 \text{ epm OH}^-$$

Then,

$$\log_{10}R = 0.0106 \times 525 - 0.17\sqrt{1.58} - 8.27$$

$$= -2.92$$

$$R = 0.0012$$

$$SiO_2 \text{ in steam} = 0.0012 \times 25 = 0.03 \text{ ppm}$$

A concentration of 0.03 ppm SiO_2 in the steam indicates that the free alkalinity is too low and suggests that the B value should be raised. This number represents the amount of free hydroxyl ion in boiler water. It is determined by adding neutral barium chloride solution to a 60-ml sample of boiler water and titrating to the end point of phenolphthalein with standard acid. P- and M-alkalinities are expressed in terms of equivalent ppm $CaCO_3$, but B values are in the unusual terms of parts of carbonate per 100,000. The B value is converted to ppm $CaCO_3$ by multiplying by 16.6; the result then compares with (2P $-$ M). In the above example the B value is

$$(2 \times 140 - 201)/16.6 = 4.7$$

The normal range of B values for 850-psi boilers is 6–8. At a B value of 8 the steam would contain 0.018 ppm SiO_2 and the pH of the boiler water would be about 11.8.

TABLE 6.19

Alkalinity of Boiler Waters

Measure of alkalinity	1	2	3	4	5	6
pH	11.4	11.4	11.4	11.4	11.4	11.4
P-Alkalinity, ppm $CaCO_3$	250.0	230.0	217.0	231.0	209.0	232.0
M-Alkalinity, ppm $CaCO_3$	280.0	263.0	256.0	262.0	232.0	267.0
(2P $-$ M), ppm $CaCO_3$	220.0	197.0	178.0	200.0	186.0	197.0
(2P $-$ M) $\times$ 0.80 = ppm NaOH	176.0	158.0	142.0	160.0	149.0	158.0
2(M $-$ P), ppm $CaCO_3$	60.0	66.0	78.0	62.0	46.0	70.0
2(M $-$ P) $\times$ 1.06 = ppm Na_2CO_3	64.0	70.0	84.0	66.0	49.0	74.0
B Value, parts CO_3^{--}/100,000	12.8	12.0	10.9	12.0	11.0	12.1
B $\times$ 13.33 = ppm NaOH	171.0	160.0	145.0	160.0	147.0	161.0

In Table 6.19 the pH, P-, M-, and B-alkalinities of several samples of water from 850-psi boilers are listed. In the samples in this table the value (2P $-$ M) is a measure of hydroxyl ion and 2(M $-$ P) is a measure of carbonate ion. It will be noted that ppm NaOH calculated from (2P $-$ M) corresponds fairly closely to that obtained from B value; this provides a check on the accuracy of the P- and M-alkalinities. Both the P and B values in Table 6.19, however, are too high. P-alkalinity should be about 80 percent of M-alkalinity, which in turn should not exceed 20 percent of the concentration of total dissolved solids—in these samples ≈ 1150 ppm. The blowdown of these boilers should be such that P-alkalinities are about 180 ppm $CaCO_3$. This would reduce B values to 7 or 8, which is sufficient to control scaling by silicates and reduce their volatility to 0.02 ppm SiO_2 in steam, while maintaining a concentration of 25 ppm SiO_2 in the boiler water. To correct the operating conditions completely, however, some changes are necessary in the treatment of the feed water. In particular, the M-alkalinity of the feed water could be reduced by adding more sulfuric acid to the make up water before passing it through the cation exchangers (see Fig. 6.3). The effect in the boilers would be to increase the ratio of total solids to total alkalinity.

c. Chemical Analysis and Control

Routine analysis in steam-generating plants serves three purposes: it makes possible the correct operation of water treating process units; it detects contamination in components of boiler feed water; it ensures adequate protection for the boiler. To accomplish these goals it is advisable to establish a testing schedule to make sure that sufficient information is obtained with the minimum number of tests. As the first step in determining the proper frequency of testing of boiler water, it is useful to calculate the length of time required for the concentration of an added chemical (e.g., phosphate) to fall to one-half its initial value, supposing none is added in the meantime. The half-depletion time, $t_{1/2}$, is calculated from Eq. (1-10) by setting $c = c_o/2$.

$$\log_e(c/c_o) = -bt/V$$

$$\log_e c_o - \log_e 2 - \log_e c_o = -bt/V$$

$$t_{1/2} = \frac{\log_e 2}{b/V}$$

$$= \frac{0.693}{b/V}$$

Thus, if $b = 771$ gph and $V = 25,000$ gal,

$$t_{1/2} = 0.693/0.0308$$
$$= 22.5 \text{ h}$$

This means that if the initial concentration of phosphate were 15 ppm, its concentration would fall to 7.5 ppm in 22.5 h, if no phosphate were added during that period of time. In the example being used, it is desirable to keep the concentration of phosphate in the range of 10–15 ppm. Therefore, in order to detect an injection pump failure or a plugged phosphate line before the chemical has fallen out of the desired range of concentration, it is obviously necessary to test for phosphate more often than once per day. Using Eq. (1-10) it can be calculated that in 8 h the concentration of phosphate would decrease from 15 to 11.7 ppm, should the supply to the boiler be interrupted. Thus, it would be advisable to measure the concentration of phosphate once every 8 h. Table 6.20 lists half-depletion times at various blowdown ratios in the 850-psi boilers being used as an example.

TABLE 6.20

Half-Depletion Times

R	Percent b	b/V	$t_{1/2}$
10	10.0	0.0926	8
15	6.7	0.0595	12
20	5.0	0.0439	16
25	4.0	0.0347	20
30	3.3	0.0287	24

Raw water must be checked occasionally for M-alkalinity and hardness distribution to determine the proper dosages for lime softening; if silica-removal is a consideration, silica concentration must also be determined to establish the proper dosage of magnesium oxide. Unless the composition of raw water is variable, one set of results per shift is sufficient. Similarly, alkalinities and hardness distribution are needed to ensure correct operation of lime-soda treaters; hot-lime softening is improved by making sure that a residual of 6–8 ppm of phosphate is maintained in the effluent. Table 6.21 is a sampling and testing schedule that provides adequate information for operating most water-treating and steam-generating plants.

TABLE 6.21

Suggested Sampling Schedule

Sample Points	P	M	(2P – M)	pH	TH	CaH	MgH	SiO₂	NaCl	Conductivity	Dissolved solids	PO₄	SO₃	Frequency, h/test
Raw water	–	x	–	–	x	x	x	x	–	–	–	–	–	8
Spaulding effluent	x	x	x	–	x	x	x	x	–	–	–	–	–	4
Lime softener effluent	x	x	x	–	x	x	x	x	–	–	–	x	–	4
Cation exchange effluent	–	–	–	x	(x)	–	–	–	–	–	–	–	–	2
Demineralizer effluent	–	–	–	(x)	–	–	–	(x)	–	(x)	–	–	–	( )
Condensate	–	–	–	x	c	–	–	c	c	(c)	–	–	–	(2)
Boiler feed water	–	–	–	(x)	x	–	–	x	x	–	–	–	x	8
Boiler blowdown water	*	x	*	*	–	–	–	x	x	x	x	x	x	8
Condensed steam	–	–	–	–	–	–	–	–	(x)	(x)	–	–	–	( )

x = Operating control test.
( ) = Continuous analyzer recommended.
c = Occasional check for contamination.
* = Control for coordinated phosphate-pH.

When sulfuric acid is added to lower the pH of lime-softened water, as in Fig. 6.3, it is advisable to check the pH of the effluent from cation exchangers to be sure the controller is functioning properly. A continuous analyzer for hardness is also convenient for monitoring the performance of cation exchangers. The instrument should be installed on the softener outlet header, then, when it alarms, the individual softeners must be checked for leakage by the versenate titration, as described in Chapter 8. In the absence of a continuous analyzer, each softener should be checked for hardness every 2 h. Continuous analyzers that monitor conductivity and pH are mandatory on demineralizers supplying feed water for high-pressure boilers. In normal operation, the pH of the cation exchange effluent is about 3, while that of the anion exchange is near 8. Should the pH of the latter be 9 or more, excessive sodium ion leakage from the cation exchange unit is indicated. As a demineralizer becomes exhausted silica appears in the effluent and its pH falls abruptly.

Neutralizing amines are usually added to condensate in an amount sufficient to raise its pH to 8-9, so the regular measurement of the pH of condensate is necessary to regulate the dosage of amines. Most contamination can be detected by a continuous conductivity monitor; if the conductivity of combined condensate rises, individual returned streams should be checked—usually for hardness and silica. On board ships, however, it is essential to measure the concentration of sodium chloride in the condensate frequently to detect condenser leaks and contamination by seawater.

Feed water for boilers being treated with chelants must be checked regularly for hardness so that the feed rate of chelant can be adjusted to the proper value. Sulfite (or hydrazine) residual should also be determined in the feed water once per shift. The concentrations of silica and chloride in feed water when compared to those in boiler water can be used to approximate the blowdown ratio or percent blowdown.

All tests listed in Table 6.21 can be performed reasonably well by operating personnel in the boiler plant. Alkalinity, hardness, chloride, and sulfite are determined by titrating with appropriate standard solutions and color indicators; silica and phosphate are determined by colorimetric methods; pH and conductivity (dissolved solids) are measured by electric meters. A number of test kits are marketed for use in boiler plants, but more reliable results are obtained by using standard laboratory equipment and glassware—burets, volumetric flasks, pipets, colorimeters, and so on. Simplified laboratory procedures are included in Chapter 8 that are no more difficult for boiler operators to use than the average test kit. A pH meter should be provided in the boiler plant. Test papers and colorimetric pH indicators are unreliable unless the pH to be measured falls in the middle of the transition range of the indicator, a circumstance that seldom occurs. Solutions of indicators have the further disadvantage of being susceptible to deterioration by sunlight, oxidizing agents, and microbiological infestation. Color and turbidity in water samples interfere and the addition of the indicator itself to unbuffered samples such as condensates or demineralizer effluent often lowers the pH of the water by one or two units. Nonroutine checks for iron, copper, oxygen, ammonia, and other contaminants had best be done in an analytical laboratory.

d. Continuous Analyzers

It is advisable to monitor the purity of steam that is used to drive turbines, by measuring either the conductivity or the concentration of sodium in a condensed sample. As steam purity is affected by certain operating variables (load swings, water level, soot-blowing, etc.) it is best to sample and monitor continuously, bearing in mind that several precautions are necessary to obtain representative samples of steam.[25] If it contains suspended particles it is essential that sampling be isokinetic, i.e., the velocity of steam entering the sampling nozzle must equal the velocity of steam in the pipe being sampled. For measuring carryover, a single-port sampling nozzle should be installed in one of the superheater supply tubes between the steam drum and the superheater. When sampling a pipeline a multiport nozzle is installed perpendicularly to the flow of steam. Sampling superheated steam, which may contain entrained dust and mists, requires a special atemperating nozzle designed so that water can be added to desuperheat the steam and thus prevent the deposition of solid contaminants on the walls of the sampling line. Also, it is preferable to sample vertical pipe runs in which the flow of steam is downward, rather than horizontal runs; bends, elbows, and valves should be avoided when installing sampling nozzles.

Saline contaminants in steam are carried into the steam header dissolved in entrained boiler water, their concentration varying to some degree with the pressure of the boiler. Below 400 psi, steam may contain up to 1 ppm of solids; at 600 psi, 0.3–0.4 ppm; in superheated steam for high-pressure turbines, the concentration of solids should not exceed 30 ppb. As a general rule, steam at pressures of 900–1200 psi is of adequate purity if it contains no more than 60 ppb of total dissolved solids, 20 ppb of sodium, and

20 ppb of silica.

The Larson-Lane steam purity analyzer* indicates carryover by measuring the conductivity of a degasified, condensed sample of steam after first passing it through a cation exchange resin in the hydrogen form. Steam is continuously condensed in a vented chamber at or near the boiling point of water, after which it passes through a conductivity cell, then through the ion exchanger where dissolved salts are converted to their corresponding mineral acids. The sample next enters a vented reboiler chamber where dissolved gases are volatilized, then through a second conductivity cell to waste. The first conductivity is a rough measure of amines, ammonia, and carbon dioxide; the second is proportional to the concentration of total dissolved solids (≈ 0.1 ppm/μmho). As the resin becomes exhausted the two conductivities converge. This instrument is not suited to steam generators in which hydrazine is used as an oxygen scavenger because so much ammonia is produced by thermal decomposition that the resin quickly becomes exhausted.

Dissolved salts are transformed to mineral acids to increase the sensitivity of the analyzer. As shown in Table 6.22, the specific conductance of a solution containing 1000 ppm of hydrochloric acid is six times that of a similar solution of sodium sulfate.

TABLE 6.22

Specific Conductances of 1000-ppm Solutions

Compound	Specific conductance (μmhos)	Conductance/ppm
$NaHCO_3$	870	0.87
Na_2SO_4	1300	1.30
NaCl	1550	1.55
Na_2CO_3	1600	1.60
NaOH	4800	4.80
HCl	7800	7.80

Steam purity can also be assessed by measuring the concentration of so-

---

* Manufactured by Beckman Instruments, 89 Commerce Road, Cedar Grove, New Jersey 07009.

dium ion, which is proportional to that of total dissolved solids (0.1 ppm $Na^+ \cong 0.3$ ppm of solids). Flame spectrophotometers have been used for this determination, but specific ion electrodes are more sensitive. The hydrogen ion activity in the sample is reduced by adding ammonia or dimethylamine, and the potential of a sodium ion electrode is compared to that of a reference electrode. Analyzers are available* that continuously measure the concentration of sodium in condensed steam at values as low as 0.1 ppb; there is no interference by hydrazine, morpholine, cyclohexylamine, or ammonia.

Another useful continuous analyzer takes advantage of the high thermal conductivity of hydrogen to monitor the concentration in steam of hydrogen gas generated in the internal corrosion of boilers.[26] A consideration of atomic weights in the reaction

$$3Fe + 4H_2O = Fe_3O_4 + 4H_2 \qquad (6\text{-}17)$$

reveals that the evolution of 1 lb of hydrogen gas corresponds to the oxidation of 21 lb of steel by hot water. Thus,

$$\text{Fe loss, lb/day} = \frac{\text{ppb } H_2 \times \text{steaming rate, lb/day} \times 21}{10^9} \qquad (6\text{-}18)$$

Small concentrations of hydrogen, normally less than 1 ppb, are also formed in the thermal decomposition of organic materials such as polymers, lignins, tannins, chelants, hydrazine, and amines. Normal levels of hydrogen, attributable to general corrosion, are 1-3 ppb, and account must be taken of this in interpreting results. Steam blanketing, impeded circulation of boiler water, flame impingement, and other sources of thermal stress increase corrosion and the rate of evolution of hydrogen.

A deaerating heater that is functioning properly delivers water that contains less than 10 ppb of dissolved oxygen, but overloading, surges in feed water demand, internal mechanical failures, or interruption of the steam supply can drastically reduce the efficiency of deaeration. For this reason, it is worthwhile installing a dissolved oxygen analyzer on the deaerator

---

* Milton Roy Company, 5000 Part St., North, St. Petersburg, Florida 33733.
  Orion Industrial, 380 Putnam Ave., Cambridge, Massachusetts 02139.

outlet. Two types of analyzers have been developed for this application.* The first of these makes use of the rapid reduction of oxygen by metallic thallium.

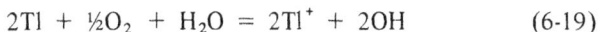

$$2Tl + \tfrac{1}{2}O_2 + H_2O = 2Tl^+ + 2OH \qquad (6\text{-}19)$$

As the result of this reaction, 1 ppm of oxygen reduced by thallium increases the conductivity of water by 35 μmhos. In order to reduce background conductivity the deaerated water is demineralized before it flows through a bed of thallium pellets, and the resulting conductivity is recorded. Thallous ion is an extremely toxic chemical, its lethal dose ranging from 10 to 30 mg/kg of body weight. In cases of poisoning it is found in all organs, although the kidney is the principal site of deposition. Symptoms of acute thallium intoxication include nausea, vomiting, pain in the extremities, convulsions, and death. Because of this toxicity, ion exchange columns are provided to remove thallous ion from the effluent of the analyzer.

The second type of instrument contains a sensing element consisting of a gold electrode connected to a silver electrode immersed in a solution of potassium chloride and isolated from the sample stream by a gas-permeable Teflon membrane. When a potential is applied to the electrodes they become polarized and no current flows. If oxygen is present, however, it permeates the membrane and is reduced at the gold cathode causing a current to flow proportional to the concentration of dissolved oxygen. The range of the instrument is 0-200 ppb O_2 and the net reaction is

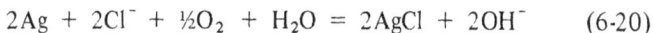

$$2Ag + 2Cl^- + \tfrac{1}{2}O_2 + H_2O = 2AgCl + 2OH^- \qquad (6\text{-}20)$$

Both conductivity and pH measurements are used to indicate breakthrough in demineralizers: as the resins become exhausted the pH falls and the conductivity rises. Then too, there are analyzers** that determine silica in demineralized water. Continuous analyzers can also be obtained** that measure the total hardness of the effluent from zeolite softeners. All of these instruments generate a signal that can be used to activate an alarm.

Lang[27] has reported the installation of continuous analyzers for phosphate, silica, and sulfite in boiler blowdown water. These would probably

---

* Beckman Instruments, 89 Commerce Road, Cedar Grove, New Jersey 07009.
** Hach Chemical Company, P.O. Box 907, Ames, Iowa 50010.

be advantageous on high-pressure boilers (1500+ psi), but would not be very useful for stationary boilers at low to moderate pressures where internal concentrations change only slowly. An instrument that measures instantaneous corrosion rates, however, does have some value in assessing the corrosiveness of boiler water. Freedman, et al.[28] have described a device for measuring electrical resistance that they used to monitor corrosion rates. The principle of the measurement is that the resistance of an electrical conductor is inversely proportional to its cross-sectional area. If a metallic sensor is inserted into the steam drum of a boiler below the water level, then, as corrosion proceeds the cross section of the metal sensor decreases and its resistance increases; the change in resistance correlates with the corrosiveness of the water. A standard method[29] is now available for measuring corrosiveness by electrical methods. The normal corrosion rate in a 600-psi boiler is 0.2–0.7 mpy; at 850 psi it is about 1 mpy, or a little less.

When considering the application of continuous analyzers, especially those that depend upon electrical sensors, it should be noted that filming amines can quickly inactivate them by forming an insulating film on the conducting surfaces.

e. Interpretation of Water Analyses

As there are a number of sources of error in analyzing boiler water, it is important to examine analytical results for self-consistency before attempting to draw any practical conclusions from water analyses. To show how to do this, it is assumed that samples from five 850-psi boilers together with their feed water are analyzed for total dissolved solids, M-alkalinity, silica, chloride, and sulfate. Further it is assumed that neutralizing amine and catalyzed sodium sulfite are added to the feed water; disodium phosphate and supplementary sodium sulfite are injected into the steam drums. Results of these analyses are displayed in Table 6.23.

In scanning the results in Table 6.23, the low value of phosphate in Boiler No. 3 immediately strikes the eye. As salts present in the feed water should all concentrate to the same extent in each boiler, it is obvious that the concentrations of silica, chloride, and sulfate should parallel each other— that is, the boiler with the highest total solids should also have the highest concentrations of silica, chloride, and sulfate. In general, these relationships seem to hold reasonably well. In order to test the results further, however, the blowdown ratios of the salts in the boilers are calculated and

TABLE 6.23

Boiler Water Analyses

Salts	Feed water	Boilers 1	2	3	4	5
Total dissolved solids (ppm)	52.0	1080	1100	1240	1320	1200
M-Alkalinity (ppm $CaCO_3$)	9.0	163	162	192	203	184
Silica (ppm SiO_2)	0.9	16	20	20	25	24
Chloride (ppm NaCl)	4.0	104	112	135	140	134
Sulfate (ppm Na_2SO_4)	22.0	400	499	591	628	572
Phosphate (ppm PO_4^{---})	–	11	11	2	13	12
Sulfite (ppm SO_3^{--})	0.5	13	14	13	17	16

compared. This is done by dividing the concentration of each salt in the water in each boiler by the corresponding concentration of each salt in the feed water. The least significant of these ratios are those of total dissolved solids, which are routinely measured by conductivity, and those of total alkalinity, which are skewed by the neutralizing amine added to the feed water. Amines contribute to the alkalinity of the feed water, but not to that of the boiler water. Table 6.24 contains the blowdown ratios of salts in the five boilers calculated from results recorded in Table 6.23.

TABLE 6.24

Blowdown Ratios

Salts	Boilers 1	2	3	4	5
Total dissolved solids	20.8	21.2	23.8	25.4	23.1
M-Alkalinity	18.1	18.0	21.3	22.6	20.4
Silica	21.1	22.2	22.2	27.8	26.7
Sodium chloride	26.0	28.0	33.8	35.0	33.5
Sodium sulfate	18.2	22.7	26.9	28.5	26.0

Ignoring for the moment the total dissolved solids and M-alkalinity ratios, it is seen that, with a couple of exceptions, the silica and sulfate ratios agree fairly well. The chloride ratios are all higher, however, suggesting that the chloride concentration in the feed water is higher than indicated in Table 6.23. As the value shown there is only 4 ppm NaCl, an error of just 1 ppm exerts a good deal of arithmetical leverage on the chloride blowdown ratios. Thus, if the concentration of NaCl in the feed water were really 5 ppm instead of 4 ppm, the blowdown ratios of the five boilers in order would be 20.8, 22.4, 27.0, 28.0, and 26.8, i.e., 20 percent lower and in good agreement with most of the silica and sulfate ratios. The sulfate value in Boiler No. 1 appears to be too low; its actual concentration is probably around 460 ppm Na_2SO_4. Also, the silica concentration in Boiler No. 3 is probably closer to 24 ppm than to the 20 ppm reported in Table 6.23. As mentioned before, the alkalinity ratios are low because neutralizing amine raises the alkalinity of the feed water. If the suggested corrections are made in the three sets of values in Table 6.23 and blowdown ratios recalculated, the results are as shown in Table 6.25. In practice, of course, the questionable results would be checked to verify these conclusions.

TABLE 6.25

Corrected Blowdown Ratios

Salts	Boilers				
	1	2	3	4	5
Silica	21.1	22.2	26.7	27.8	26.7
Sodium chloride	20.8	22.4	27.0	28.0	26.8
Sodium sulfate	21.0	22.7	26.9	28.5	26.0
Average	21.0	22.4	26.9	28.1	26.5

Returning now to the low value of phosphate in Boiler No. 3, this result should, of course, be checked before drawing any conclusions. It is seen in Table 6.25, however, that the average blowdown ratio in Boiler No. 3 is 26.9, and in Table 6.23 that the feed water contains 0.5 ppm of sulfite. Therefore, the concentration of sulfite in Boiler No. 3 attributable to that in the feed water is $26.9 \times 0.5 = 13.5$ ppm. As an additional 3 or 4 ppm of sulfite should be introduced with the phosphate, it can be concluded that either the phosphate injection line to Boiler No. 3 is plugged or the chemical injection pump is not working. Incidentally, the blowdown on Boilers No. 1 and No. 2 can be decreased to bring the silica concentrations up to their limiting value of 25 ppm SiO_2.

Samples of boiler water are usually taken from the continuous blowdown line before flashing. Referring to Fig. 1.1, it is apparent that the water near the blowdown outlet in the steam drum contains higher concentrations of salts than water entering the downcomers, which is diluted by the entering feed water. In addition, concentrations of dissolved salts vary with the steaming rate. Gallatin and Dollison,[30] in a detailed analytical study of boiler water at varying loads and pressures, found that the concentrations of silica and phosphate were inversely proportional to steaming rates and that the rate of change of steaming rate also affected analytical results significantly. It has further been observed that whenever a boiler is brought onto the steam header, the concentration of silica in both the boiler water and in the steam increases temporarily, then returns to normal.[31] The sensitivity of concentration to steaming rate may be attributable to the following mechanism. When a bubble of steam escapes from a heating surface, it leaves behind a concentrated film of boiler water from which certain of the less-soluble components (most likely silica, sulfate, and phosphate) may precipitate. If the steaming rate is so high that washing of the heat transfer surface by boiler water is inadequate, these otherwise-soluble salts may remain out of solution on the surface. Upon reducing the steam load some dilution of the boiler water occurs as a smaller percentage of the total volume of the boiler is occupied by steam. Steam generating surfaces then become more accessible to normal flushing by the recirculating water, and the precipitated salts redissolve. This phenomenon is commonly called "hideout."

Before making use of a water analysis for operating purposes, the careful plant manager will wish to verify its integrity, particularly if it is represented to him as a complete analysis. A fundamental principle of the chemistry of solutions is that in any solution the equivalents of positive and negative ions are equal. Therefore, the accuracy and completeness of a water analysis can be checked by converting the conventional units to equivalents per million and determining whether or not the concentration of total cations equals that of total anions. Alkalinities and hardness are customarily reported in terms of ppm $CaCO_3$, which is converted to ppm of ions by multiplying by the ratio of the equivalent weight of the ion to the equivalent weight of calcium carbonate. To convert ppm of ion to epm, one must multiply by the reciprocal of the equivalent weight of the ion. To illustrate the procedure, the analysis of raw water given in Table 6.14 will be verified. Alkalinity relationships determined by P- and M-alkalinity can be found in Table 5.4. Factors are given in Table 6.26 for converting ppm $CaCO_3$ to ppm ion and to epm ion.

TABLE 6.26

Conversion of Concentration Units

(TH – CaH)	X	0.24	=	ppm Mg^{++}	X	0.082	=	epm Mg^{++}
CaH	X	0.40	=	ppm Ca^{++}	X	0.050	=	epm Ca^{++}
(M – 2P)	X	1.22	=	ppm HCO_3^-	X	0.016	=	epm HCO_3^-
2P	X	0.60	=	ppm CO_3^{--}	X	0.033	=	epm CO_3^{--}
				ppm Na^+	X	0.043	=	epm Na^+
				ppm Cl^-	X	0.028	=	epm Cl^-
				ppm SO_4^{--}	X	0.021	=	epm SO_4^{--}
ppm SiO_2	X	1.27	=	ppm SiO_3^{--}	X	0.026	=	epm SiO_3^{--}

The analysis in Table 6.14 can be verified by substituting the values given there in Table 6.26. The results are summarized in Table 6.27.

TABLE 6.27

Verification of a Water Analysis

Original analysis (ppm)		Ions	Cations (ppm)	Anions (ppm)	Cations (epm)	Anions (epm)
TH	79	Mg^{++}	7.7		0.64	
CaH	47	Ca^{++}	18.8		0.94	
Na	15	Na^+	15.0		0.65	
P	4	HCO_3^-		69.5		1.11
M	65	CO_3^{--}		4.8		0.16
Cl	7	Cl^-		7.0		0.20
SO_4	17	SO_4^{--}		17.0		0.36
SiO_2	12	SiO_3^{--}		15.0		0.40
TDS	160	Totals:	41.5	113.3	2.23	2.23

As the total equivalents per million of anions equals that of the cations it can be assumed that the analysis is correct as given and that nothing has been left out. Note that although the equivalents per million are identical, the parts per million of cations and anions are quite different. It is possible for them to be equal, but to find this in a natural water is highly unlikely.

6.3 OPERATING GUIDELINES

The chapter is concluded with some miscellaneous information having to do with practical operations of boiler plants: for convenience these items are classified as chemical and mechanical. Among the chemical topics are procedures for laying-up boilers, proper methods of treatment to prepare a boiler for shutdown, and costs of water and chemical treatments. Mechanical aspects include energy conservation, steam quality, desuperheating, and some special procedures relevant to the operation of marine boilers.

a. Chemical Considerations

For any given water supply the cost of chemical treatment is easily determined. When more than one source of supply is available, however, the determination of minimum treating cost is more complicated. Suppose, for instance, that it is necessary to generate 100,000 lb of process steam per hour in a 300-psi boiler, using 100 percent make up, the only external water treatment being filtering and softening by cation exchange. Internal treatment is to be accomplished with a single chemical formulation priced at $0.80/lb and applied at a dosage of 50 ppm. The daily cost of internal treatment depends upon the blowdown rate of the boiler and is calculated by the following formula.

$$\$/\text{day} = 50\,(\text{ppm}) \times 0.80\,(\$/\text{lb}) \times 8.34\,(\text{lb water/gal}) \times$$
$$\text{blowdown rate (gpd)} \times 10^{-6}$$

The blowdown rate, of course, is governed by the quality of the feed water. To illustrate this, suppose two sources of water, I and II, are available with compositions as shown in Table 6.28.

TABLE 6.28

Water Compositions

Components	Source of water I	II	Limitations, 300-psi boiler, maximum
Silica (ppm SiO_2)	12	2	175
M-alkalinity (ppm $CaCO_3$)	25	125	700
Total dissolved solids (ppm)	70	170	3500

The data in Table 6.28, when plotted as shown in Fig. 6.4, readily gives the proper mixture of waters I and II for minimum blowdown. The figure is constructed by plotting the percentages of I in II against the minimum blowdown necessary to maintain the limits specified in the last column in Table 6.28 (also see Table 6.12). To plot the M-alkalinity line, for example, it is noted that at zero percent of I (100 percent of II), the blowdown ratio is $700/125 = 5.6$, and percent blowdown is $100/5.6 = 17.9$. At 100 percent of I the corresponding blowdown ratio is $700/25 = 28$, and percent blowdown is $100/28 = 3.6$. Silica and total dissolved solids lines are plotted similarly. It is apparent from an inspection of Fig. 6.4 that the proper mixture for minimum blowdown is 83 percent I and 17 percent II. At less than 83 percent of I more than 6 percent of blowdown is needed to keep the M-alkalinity at a maximum of 700 ppm $CaCO_3$. At more than 83 percent I more than 6 percent of blowdown is required to keep the concentration of silica at a maximum of 175 ppm SiO_2. Total dissolved solids can never be the limiting factor because of the relative positions of the three curves. In operating any boiler limiting concentrations may vary from time to time, with blowdown being controlled by whichever component reaches its limiting value first.

If, as is often the case, water from one source costs more than that from another, the determination of minimum cost is a little more difficult. For example, if I costs $0.45/1000 gal and II costs $0.28/1000 gal, it is necessary to compare water costs of various combinations of the two sources with the corresponding chemical costs. This is done in Table 6.29, where it is seen that despite the highest cost of chemical treatment, the total cost using II by itself is less than any combination of the two sources.

TABLE 6.29

Water Treating Costs

% I	% II	Water cost ($/1000 gal)	Minimum blowdown (%)	Feed (gpd)	Blowdown (gpd)	Chemicals ($/day)	Water ($/day)	Total ($/day)
0	100	0.28	18	350,327	62,558	20.87	98.09	118.96
25	75	0.32	14	334,944	47,175	15.74	108.02	123.76
50	50	0.37	11	323,296	35,527	11.85	119.62	131.47
75	25	0.41	7	309,429	21,660	7.22	126.86	134.08
83	17	0.42	6	306,137	18,368	6.13	128.58	134.71
100	0	0.45	7	309,429	21,660	7.22	139.24	146.46

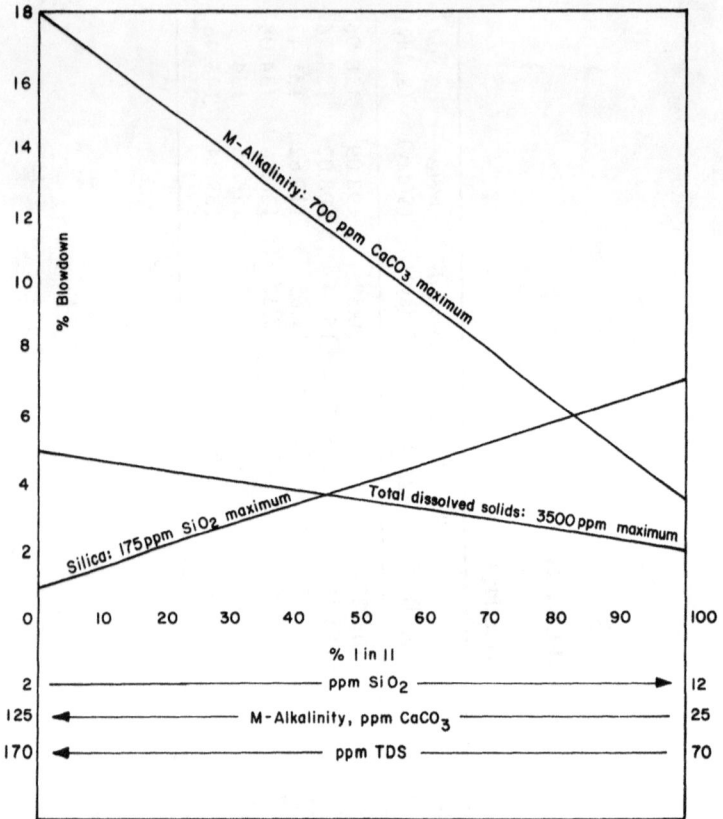

Fig. 6.4

Except in high-pressure boilers, water tubes are inserted in holes drilled in the drums and headers and then expanded to form a strong, tight seal; the process is called rolling. Occasionally, roll leaks develop, notably after replacing a phosphate treatment with a chelant program. Phosphate is quite effective in stopping small leaks and, in fact, experienced boiler operators, as an expedient, often seal leaks by temporarily increasing the phosphate feed to the boiler. Up to 123 C (253 F) the solubility of trisodium phosphate increases as temperature increases.[32] Above this temperature the solubility decreases, reaching a minimum of approximately 15 ppm at about 325 C (617 F). The precipitation of excess Na_3PO_4 seals leaks.

In Chapter 5, pages 78–81, the chemistry of hydrazine as an oxygen scavenger was discussed. Table 6.30 contains results of tests made on a ship to compare the activities of catalyzed and uncatalyzed hydrazine at the lower feed water preheater temperatures that prevail while a vessel is slow-steaming. The fourth stage heater establishes the temperature of feed water just ahead of the economizer—the zone in which oxygen scavenging should ideally be complete. In this arrangement, heat is supplied by extraction steam at 300 psi from the main turbine giving, at full power, a temperature of 410 F at the outlet of the fourth-stage heater. As power is reduced the pressure of the extraction steam falls until, at around 225 psi (390 F at the heater outlet), automatic switch-over occurs and let-down steam is supplied in place of extraction steam. Tests were also done at 375 F using let-down steam.

TABLE 6.30

Qualitative Effectiveness of Hydrazine

Type of hydrazine	Hydrazine residual (ppm)	4th-Stage heater outlet		
		410 F	390 F	375 F
Catalyzed	0.01	s	s	s
Uncatalyzed	0.01	m	m	u
Catalyzed	0.03	s	s	s
Uncatalyzed	0.03	s	s	m
Catalyzed	0.08	s	s	s
Uncatalyzed	0.08	s	s	m

s = satisfactory; m = marginal; u = unsatisfactory

These results show that a 0.03-ppm residual of hydrazine in the feed water gives satisfactory oxygen scavenging with either catalyzed or uncatalyzed hydrazine, and that the outlet temperature of the fourth-stage heater should not be allowed to fall much below 390 F if satisfactory scavenging is to be achieved with uncatalyzed hydrazine.

When bringing a boiler down for its annual inspection or for major repairs it is desirable that the internal surfaces of the boiler be as clean as possible. If the shutdown is scheduled, it is advisable to double the normal blowdown rate for the two weeks preceding shutdown, at the same time increasing the dosage of sludge conditioner to about one and one-half times normal. Also, the boiler should not be drained on shutdown until it is cool.

If a boiler is to be placed on standby and is being kept filled, ready for service on short notice, it is well to maintain a blanket of nitrogen in the vapor spaces of the boiler and superheater to prevent corrosion of the wet metal. Ordinarily, the boiler is placed on standby directly as it comes off the line. Sufficient caustic is added to give an M-alkalinity of 500 ppm $CaCO_3$; the concentration of sodium sulfite is raised to 150-200 ppm. Next, a nitrogen cylinder equipped with a 2-stage pressure regulator is connected to valves on the steam drum and superheater vents. When the steam pressure falls to 5-10 psi, set the nitrogen pressure at 10-15 psi, then open the valves to the boiler and superheater. Maintain a positive pressure of 5 psi of nitrogen while the boiler cools. This method avoids trying to flood the superheater—which is often impossible—and also the formation of salt deposits in it when the boiler is returned to service.

Hydrazine, which is alkaline and forms no solids, is useful for laying up boilers that have nondrainable superheaters. In this case the boiler should be completely filled with water containing 300-350 ppm of active hydrazine prepared by mixing concentrated hydrazine with hot condensate containing a low concentration of oxygen.

b. Mechanical Considerations

This section has very little to do with chemical treatment, but is included to explore, cursorily, some important mechanical aspects of the operation of boilers and water-treating facilities. Fig. 6.5 shows some possible interconnections in a steam plant producing superheated steam at 850 psi. A combination extraction and condensing turbine is driven by the high-pressure steam, part of which is extracted at 125 psi to provide a supply of process steam. Supplementary sources of 125-psi steam are a desuperheating station and a flash tank. Blowdown water from the steam drum enters the high-pressure flash tank where steam is flashed at 125 psi. Water from this vessel flows to the low-pressure flash tank producing 15-psi steam, which serves the deaerating heater shown in Fig. 6.3; the remaining hot water goes to the feed water preheater. In operating an extraction-condensing turbine the flow to the condenser through the extraction port should be minimized, as heat is otherwise wasted through the condenser to the atmosphere. The back-pressure section of the turbine should be used to supply as much as possible of the total steam, even though this reduces the total energy available to the machine.

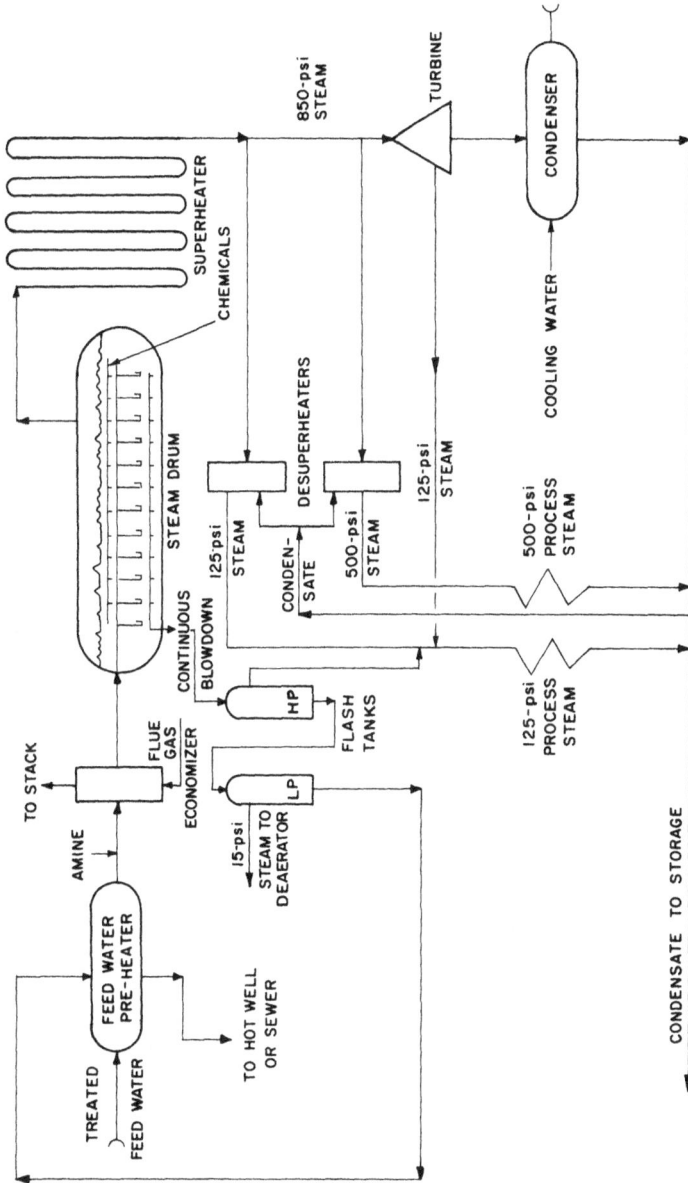

Fig. 6.5 Moderate pressure steam plant.

As process steam is considered a utility and an operating necessity, it is therefore an expense. For this reason, efforts are directed toward producing steam at maximum efficiency to reduce the unit cost to consumers. Boiler efficiency (steam-to-fuel) is defined as the ratio of total energy in the steam generated to the energy in the fuel consumed in producing the steam.

$$\text{Efficiency} = \frac{\text{pounds of steam} \times \text{enthalpy} \times 100}{\text{fuel consumed} \times \text{high heating value}} \qquad (6\text{-}21)$$

The term high heating value refers to the gross heat of combustion of a fuel, as determined in an adiabatic calorimeter, in which water vapor formed in the combusiton of the fuel is condensed.[33] Low heating value is the net heat of combustion realized in a furnace, where the water vapor is not condensed, but goes up the stack. The difference is essentially the latent heat of evaporation of water, which is recovered in a calorimeter but not in a furnace. The price of fuel is customarily based on its high heating value.

The temperature and pressure of the feed water must also be known to determine the amount of heat actually added to generate steam. Energy losses inevitably occur, of course, principally in the boiler blowdown, stack gas, and by conduction and radiation of heat. By using flash tanks, as in Fig. 6.5, a great deal of heat is recovered from the hot blowdown water, but if the water is flashed to the atmosphere in a hot well, a large proportion of its heat escapes. It is instructive to compute the heat loss in equivalent barrels of fuel oil (EFOB), defined as 6.3×10^6 Btu/EFOB. The heat lost is that required to heat boiler feed water to the temperature of the blowdown water, i.e., the difference in enthalpy (Btu/lb) of the two streams. These values can be found in steam tables.[34] As an example, suppose 5000 lb/h of blowdown water is being withdrawn from a 450-psi boiler and flashed to atmospheric pressure. The temperature of the feed water is 125 F and the efficiency of the furnace is estimated as 80 percent. From steam tables:

Enthalpy of water at 125 F =   93 Btu/lb

Enthalpy of water at 456 F = 436 Btu/lb

Therefore, the heat needed to raise 5000 lb/h of water from 125 to 456 F is

$$Q = (436 - 93)(5000)/0.80 = 2,100,000 \text{ Btu/h}$$

Dividing this heat by 6.3×10^6 Btu/EFOB gives

$$\frac{2,100,000}{6.3 \times 10^6} = 0.33 \text{ EFOB/h}$$

or 8 barrels of fuel oil per day.

In the interest of economy, several things can be done to improve the efficiency of a boiler. The temperature of the feed water can be raised by condensing exhaust steam in a deaerating heater, as in Fig. 6.3; by passing the feed water through an economizer in the convection section of the furnace; or by heat exchange with a hot stream in a nearby process plant. Another means of conserving heat is to preheat combustion air with the hot gases passing from the furnace to the stack. An exceptionally efficient air preheater is the regenerative type in which a rotating metal drum, made up of honeycomb sections, revolves through the flue gas duct absorbing heat, then through the incoming air duct surrendering heat.

Several variables are important in ensuring proper combustion in the furnace. These include the amounts of fuel and air to the furnace, the draft or differential pressure across the furnace, and the temperature of the exiting flue gas. Proper control of the fuel/air ratio ensures a high concentration of carbon dioxide in the flue gas, which in turn raises the temperature of the furnace. Air in excess wastes heat to the atmosphere and reduces efficiency by absorbing heat otherwise available for generating steam. The theoretical requirement of fuel oil for air is 18-20 lb air/lb fuel oil. Glaubitz[35] reports favorable results with only 1-2 percent of excess air in combating ash deposition and low-temperature corrosion. Less sulfur trioxide is formed when excess air is reduced. Attig and Sedor[36] corroborate this, but point out that unbalanced fuel/air ratios can create a reducing atmosphere and form sulfides, which are themselves intensely corrosive to ferrous metals. In any event, to achieve optimum efficiency matters must be so arranged that combustion is virtually complete in the furnace.

Table 6.31 displays results obtained with a small gas heater used to warm water flowing at about 30 gal/h with an initial temperature of 52 F. In this demonstration device the maximum furnace temperature and highest rate of heat transfer were obtained at 3 percent of oxygen, or 15 percent excess air.

Combustible gas analyzers detect unburned fuel by means of an elec-

TABLE 6.31

Combustion Efficiency

% Oxygen	% Combustibles	Temperature (F) Furnace	Stack	Water	Heat transfer (Btu/h)
15.0	0	936	615	92	10,000
11.0	0	1083	650	99	11,700
6.0	0	1310	675	103	12,800
4.0	0	1340	670	106	13,500
3.0	0	1345	660	107	13,800
1.5	0.2	1310	650	105	13,300
1.0	3.1	1235	600	101	12,200

trical resistance element or thermistor coated with an oxidation catalyst and installed in a bridge circuit. Any combustible material in a sample of flue gas is oxidized on the catalyst, raising the temperature of the element and increasing its resistance. This unbalances the bridge circuit, the degree of imbalance being proportional to the concentration of combustible material. Gilbert[37] has described a much more precise method for controlling combustion based on the determination of carbon monoxide, formed by incomplete oxidation, in the flue gas. As excess air to a burner is reduced, approaching the minimum amount necessary for complete combustion, the first detectable effect is an increase in the concentration of CO. In the presence of an adequate excess of air the level of CO is 20-50 ppm, but when the excess air falls to about 5 percent (1 percent oxygen), the concentration of CO rises abruptly by several hundred parts per million. Thus, a continuous carbon monoxide analyzer in combination with an oxygen analyzer is very useful for setting realistically low values for excess oxygen. This value varies with the fuel used and also from furnace to furnace.

Natural gas—the fuel of choice, if available—is distributed at about 60 psi, then, in most plants, its pressure is reduced to approximately 35 psi in the burner supply header. The flow of gas to the furnace is controlled automatically by the steam pressure in the boiler. A drop in steam pressure causes the control valve to open, increasing the flow of fuel; an increase in steam pressure causes the valve to pinch back, decreasing the fuel to the burners. Natural gas contains some 22,000 Btu/lb and, as it burns with a nonluminous flame, most of its heat is transferred by convection after the gas leaves the furnace proper. The temperature of the flame is around 3300 F with 10-15 percent excess air.

When burning fuel oil atomizing steam is used to break up the stream of fuel oil into very small droplets and facilitate combustion. Fuel oil burners are supplied through individual regulators that lower the supply pressure to the optimum pressure for the burner. Fuel oil averages 20,000 Btu/lb, with a flame temperature of 3400 F at 10-20 percent of excess air; a large proportion of its energy is radiant. Pulverized coal, also burned as a fluid fuel with 10-20 percent excess air, gives flame temperatures of 2000-3000 F. It is an incandescent fuel, producing a great deal of radiant energy.

In order to keep the stack temperature down it is necessary to avoid excessive air and to keep the boiler tubes clean. Most boilers are equipped with soot blowers that discharge steam within the furnace to dislodge slag, loose ash, soot, and other particulates from heat transfer surfaces. When soot is blowing with steam, carryover increases in the boiler because of turbulence caused by the drop in internal pressure. Foster[38] has pointed out the advantages of using compressed air at 250-350 psi instead of steam in large steam generators with high soot-blowing requirements. As a furnace becomes increasingly dirty the load on the boiler may have to be reduced to avoid exceeding the maximum limitation on stack gas temperature, usually 900 F. For this reason, fan blades and air preheaters should be washed once a month to remove the large amounts of iron oxides and salts that deposit there, especially when burning fuel oil containing high concentrations of sulfur.[39] Ideally, the temperature of the flue gas should be 50-75 F above the dew point—if sulfuric acid is present, about 375 F.

A most important characteristic of steam is its moisture content. Wet steam can damage machinery, cause deposits in superheaters and turbines, and in extreme cases, water hammer can break valves and fittings. The quality of steam, as measured by the moisture it contains, deteriorates under a number of operating conditions. Improper distribution of fires results in uneven heating that can cause priming, in which boiler water is thrown into the steam outlet. Load swings—the rule rather than the exception in many plants generating process steam—have a similar effect, as do high or low water levels in the steam drum. If the water level is too low, the steam separation surface is diminished and water may be thrown into the steam system by the intense turbulence. Also, if the water level is below the openings in the blowdown line, the concentration of total dissolved solids will increase, initiating foaming in the boiler. Actuating soot blowers or intermittent blowdown valves also produces turbulence in the boiler. In marine boilers the roll of the vessel in inclement weather sometimes throws water into the steam separators. All of these conditions having ill effects on steam quality can be eliminated or minimized by paying attention to operating procedures.

It is advisable to measure the quality of steam regularly with a throttling calorimeter. This instrument receives steam from a sampling nozzle attached through a pipe nipple to a valve, the outlet of which leads to a throttling plug. The latter contains an orifice sized to permit the passage of approximately 100 lb of steam per hour into an insulated chamber, then to the atmosphere. The chamber is so designed that escaping steam surrounds it, ensuring a uniform internal temperature of 212 F. (This temperature varies slightly, of course, with the pressure of the atmosphere.) Dry saturated steam, when expanded through an orifice to lower pressure, becomes superheated. If it contains no moisture the degree of superheat is registered on a thermometer under the throttling plug; if moisture is present it is converted to saturated steam by the superheat. Excessive moisture, however, dissipates all superheat and the resulting steam will be at 212 F. The percentage of moisture is calculated using the following formula.

$$\text{Percent } H_2O = \frac{[h_1 - h_2 - c(t_s - t_c)]\,100}{L} \qquad (6\text{-}22)$$

where

h_1 = enthalpy of saturated vapor to calorimeter (Btu/lb)
h_2 = enthalpy of vapor at atmospheric pressure (Btu/lb)
c = specific heat of saturated steam (Btu/lb F)
L = latent heat of vaporization (Btu/lb)
t_s = temperature of superheat in calorimeter (F)
t_c = temperature in calorimeter (F)

The values of h_1, h_2, c, and L are obtained from steam tables[34] using absolute pressures. The term $c(t_s - t_c)$ is the sensible heat in the superheated steam. As an example of the calculation, assume that the determination of the quality of steam being produced in an 825-psi boiler is desired. It is found that the superheat temperature in the calorimeter, t_s, is 305 F and the atmospheric pressure, as measured by a barometer, is 14.7 psi.

From steam tables:

h_1 = 1198.0 Btu/lb at 840 psia
h_2 = 1150.5 Btu/lb at 14.7 psia
c = 0.472 Btu/lb F at 840 psia
L = 681.5 Btu/lb at 840 psia
t_s = 305 F
t_c = 212 F at 14.7 psia

Substituting in Eq. (6-22):

$$\text{Percent } H_2O = \frac{[1198.0 - 1150.5 - 0.472(305 - 212)]\,100}{681.5}$$

$$= 0.5$$

The steam quality is thus, $(100 - 0.5) = 99.5$ percent.

The steam plant shown schematically in Fig. 6.5 has two desuperheating stations (also called atemperators) where water is sprayed into 850-psi steam to produce process steam at 500 psi and 125 psi. Fig. 6.6 shows the details of operation of the 125-psi station. A rise in pressure in the 125-psi steam header increases the pressure on the diaphragm of the master controller, which decreases the water pressure applied to the diaphragm of the pressure reducing valve. The pressure reducing valve closes slightly and the decreased flow of steam to the desuperheater causes a drop in temperature of the steam leaving the desuperheater. This temperature drop causes the temperature controller to increase the water pressure on the diaphragm of the water supply valve. It closes slightly, reducing the flow of water to the desuperheater, thus restoring to normal the temperature of the steam leaving the desuperheater. A fall in pressure in the 125-psi steam header causes opposite responses that increase the flow of steam to the header. Failure of the water supply to the desuperheaters closes them off, for the system is controlled by water pressure. As the concentration of total dissolved solids in the spray water should be < 2.5 ppm, condensate is the preferred supply. In emergencies it may be necessary to use other sources, however, in which case feed water, treated make up, or as a last resort, raw water are used. When these alternative sources are used continuously there is a possibility in some circumstances of caustic embrittlement developing in the steam line following the desuperheater; this is discussed further in Chapter 7.

This chapter is concluded with a few remarks relating to marine operations. All usable water aboard ship is distilled from seawater in evaporators, so it is important that they be operated carefully. Anything that

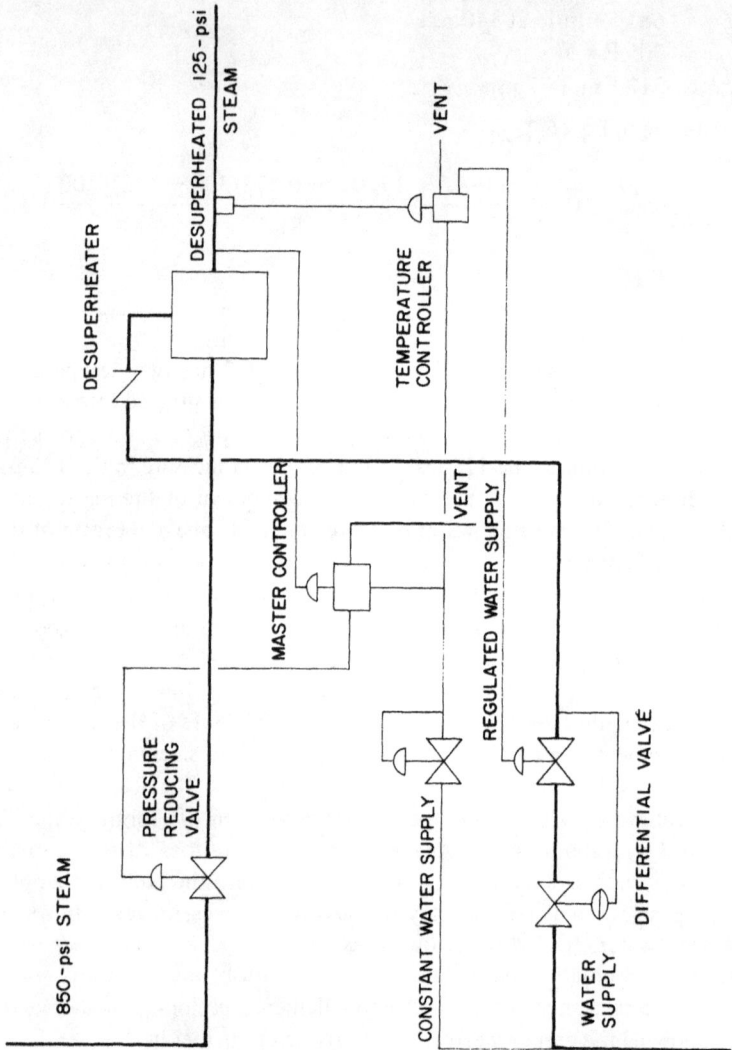

Fig. 6.6   Desuperheater station.

impairs the performance of the evaporators can create serious upsets in the boilers and can also affect the health and comfort of the crew. The most common difficulty experienced with evaporators is the formation of scale on the tubes; a relatively thin layer, 1/16- to 1/18-in. thick, can cause a reduction of 20-25 percent in the output of distilled water. As the concentration of calcium and magnesium salts increases in the brine, crystalline solids begin to deposit on the hotter tube surfaces where, because of local evaporation, the concentration of salts is much higher than in the bulk of the solution. Scaling and depositing of these crystalline compounds can be controlled in two ways: 1. An impurity can be introduced into the crystals as they form that either blocks further growth, or introduces strain into the crystalline structure; 2. Ions can be added that are adsorbed at the surface of the crystal, slowing and otherwise interfering with its further growth. The adsorption of ions on the surface of the crystal is the most important factor in controlling the growth process, but the rate is also strongly influenced by the initial supersaturation of the solution.

Two types of chemicals are used to inhibit scaling in evaporators—polyphosphates and polymers. Polymers, in general, are more satisfactory than the polyphosphates because they do not hydrolyze or lose their effectiveness. If the brine solution is allowed to concentrate excessively, however, polymers cannot function properly. Seawater contains about 1 lb of salt in 32 lb of water: this is commonly referred to as a density of "one thirty-second." The density of the brine should not be allowed to exceed 1.5 thirty-seconds, so to avoid exceeding this limitation, approximately 2 tons of brine must be pumped overboard for each ton of water distilled. Failure to limit brine concentration results in the formation of a thick, soft, spongy mass that blankets the tubes. Although this material can easily be washed off with a hose, it is an effective insulator that markedly reduces the rate of evaporation. Careful control of brine density and polymer dosage keeps most evaporators clean for many months, but to achieve good thermal efficiency and to minimize scaling it is also essential that low-pressure bleed steam at 12-15 psi be used for heating the equipment, rather than high-pressure reserve steam.

Evaporators should not be operated while the ship is in a harbor where bacterial contamination may be present, for such contaminated waters constitute a health hazard if bacteria are carried over with the distillate and thus introduced into the potable water system. Potable water should be obtained from shoreside facilities if the potable tanks are low on arrival.

Operation in shallow, sandy areas can introduce large amounts of silica into evaporators, some portion of which can be expected to carry over into the feed water. As the tolerance of high-pressure boilers for silica is low, operation in these areas should be avoided, if possible. Also, if the evaporators are in operation as the ship moves into a shallow harbor, the high-salinity alarm may sound when, in fact, the salinity level is normal. Harbor waters often contain high concentrations of organic materials and dissolved gases, which frequently increase the conductivity enough to activate the alarm.

Foaming in evaporators is another source of contamination in the distillate. Usual causes are over-concentration of brine, operation at rates much higher than design, and contamination, especially oil, entering with the seawater. To eliminate this nuisance most polymer formulations for treating evaporators contain antifoaming agents.

The blowdown of marine boilers is managed differently from that of stationary boilers. Continuous blowdown is either to the bilges or to the shell of the evaporator, often through a conductivity monitor on the water in the steam drum, the rate of flow being controlled by a 1/16-in. orifice in the line. Because of safety restrictions this rate is so low that relatively dense suspended solids do not move into the continuous blowdown line, but instead concentrate in the boiler. For this reason, high-volume blows are made intermittently to the bilges or overboard from the surface, bottom, and water wall blowdown valves on the boilers. This is done every 7–14 days, whether the vessel is in port or at sea. Thus, if a voyage is expected to last 10–14 days, blowdown should be done just before departure and then again in the middle of the voyage. Similarly, if the vessel is to remain in port longer than three days, the boilers should be blown down upon arrival and again before departure. When boilers are under exceptionally good control, the period between high-volume blowdowns can be extended to three weeks, in which event blowdown can usually be done in port. If it is done at sea, the following procedure is used:

1. Slow the engines and level off the plant for one-boiler operation. Secure fires in the boiler to be blown down, and open the superheater vent.
2. Allow the pressure to fall to at least 15 psi below the normal boiler pressure, at which level there is no danger of interrupting circulation.
3. Open blowdown valves wide, then close them in a single operation, taking 15–20 s to open and close the valves. Maintaining a visible

water level at all times, open and close the blowdown valves in the
following sequence:
   a)   Surface blow on the steam drum.
   b)   Bottom blowdown on the mud drum.
   c)   Side water wall header.
   d)   Rear water wall header.
   e)   Front water wall header.
4.   When blowdown is complete, restore the water level to normal, light
     off the burners, and bring the pressure up to normal. Put the boiler
     back on the line and close the superheater vent.
5.   Secure the other boiler and repeat the foregoing procedure.
6.   Treat the boiler with chemicals, test, and adjust as necessary to re-
     establish concentrations within limits specified for the boilers.

The surface blow mentioned in step 3a) is used to remove oil, grease,
and scums; its outlet is located 1/2-in. below the normal water level in the
steam drum.

REFERENCES

(1) Klein, H. A., Kurper, J. J., and Schuetzenduebel, W. G. 1965. Cycle cleanup for supercritical-pressure units. *Proc. Amer. Power Conf.* 27:756.

(2) Shields, C. D. 1961. *Boilers: types, characteristics, and functions.* New York: McGraw-Hill.

(3) Wangerin, D. D. 1964. Waste-heat boilers—principles and applications. *Proc. Amer. Power Conf.* 26:682.

(4) Skrotzki, B. G. A. Sept., 1958. Fuels: a look ahead. *Power* 102:75.

(5) Denman, W. L. and Salutsky, M. C. Sept., 1968. Boiler scale control. *Power* 112:80.

(6) Metcalf, J. R. Jan.,1971. Boiler chelant treatment: an update. *Ind. Water Eng.* 8 (1):16.

(7) Walker, J. L. and Stephens, J. R. 1973. A comparative study of chelating agents: their ability to prevent deposits in industrial boilers. *Proc. Intl. Water Conf.* 34:134.

(8) Meites, L., ed. *Handbook of analytical chemistry.* 1963. New York: McGraw-Hill. pp. 1–45.

(9) Venezky, D. L. 1971. Thermal stability of EDTA and its salts. *Proc. Intl. Water Conf.* 32:37.

(10) Venezky, D. L. and Moniz, W. B. *Thermal stability of nitrilotriacetic acid and its salts in aqueous solutions.* U.S. Clearinghouse Fed. Sci. Tech. Inform. AD 1970, No. 715776, 13 pp.

(10a) Swanson, D. A. Dec. 1967. Advances in boiler water treatment. *Ind. Water Eng.* 4(12):22.

(11) Edward, J. C. and Merriman, W. R. 1963. Use of chelating agents for continuous internal treatment of high pressure boilers. *Proc. Intl. Water Conf.* 24:35.

(12) Lorenc, W. F. and Bermer, R. A. May 30, 1972. Compositions and methods for controlling scale. U.S. Patent No. 3,666,664.

(13) Ralston, P. H. 1969. Scale control with aminomethylenephosphonate. *J. Petro. Techn.* 21:1029.

(14) Klein, H. A. Oct., 1962. Use of coordinated phosphate treatment to prevent caustic corrosion in high pressure boilers. *Combustion* 34 (4):45.

(15) Noll, D. E. 1964. Factors that determine treatment for high-pressure boilers. *Proc. Amer. Power Conf.* 26:753.

(16) Pocock, F. J. and Stewart, J. F. Jan., 1963. The solubility of copper and its oxides in supercritical steam. *Trans. ASME. J. Engineering Power* 85-A:33.

(17) Pocock, F. J., Lux, J. A., and Seibel, R. W. 1966. Control of iron pickup in cycles utilizing carbon steel feedwater heaters. *Proc. Amer. Power Conf.* 28:758.

(18) Duff, J. H. and Levendusky, J. A. 1962. Powdex—a new approach in condensate purification. *Proc. Amer. Power Conf.* 24:739.

(19) Grant, J. S. and Crouse, R. P. 1966. History of powdex condensate polishing equipment at bay shore station. *Proc Amer. Power Conf.* 28:773.

(20) Brindisi, P. Apr., 1959. Oxygen and carbon dioxide corrosion. *Combustion* 30:47.

(21) Pocock, F. J., Lux, J. A., and Seibel, R. W. 1966. Control of iron pickup in cycles utilizing carbon steel feedwater heaters. *Proc. Amer. Power Conf.* 28:758.

(22) Bonafede, G. and Sandell, J. W. Dec., 1965. Copper, iron, and nickel pickup in a power station feed water system. *Australasian Corrosion Eng.* 9:19.

(23) Barker, P. A. Mar.-Apr., 1975. Water treatment for steam generating systems. *Ind. Water Eng.* 12 (2):5.

(24) Coulter, E. E., Pirsch, E. A., and Wagner, E. J. Jr. 1956. Selective silica carry-over in steam. *Trans. ASME* 78:869.

(25) Amer. Soc. Testing Materials. 1977 Annual Book of ASTM Standards. Part 31, D1066-69: 60. *Standard method of sampling steam.*

(26) Jacklin, C. and Wiltsey, D. G. May, 1971. Dissolved hydrogen analyzer—a tool for boiler corrosion studies. *Materials Protection and Performance* 10 (5):39.

(27) Lang, M. Apr., 1968. Automatic water sampling and analysis. *Power Eng.* 72:45.

(28) Freedman, A. J., Troscinski, E. S., and Dravnieks, A. 1958. An electrical resistance method of corrosion monitoring in refinery equipment. *Corrosion* 14:175t.

(29) Amer. Soc. Testing Materials. Philadelphia, Pa. 1977 Annual Book of ASTM Standards, Part 31, D2776-72. *Standard test methods for corrosivity of water in the absence of heat transfer (electrical method).*

(30) Gallatin, J. C., Jr. and Dollison, J. H. 1970. Effects of load variations on boiler water constituents. *Proc. Amer. Power Conf.* 32:729.

(31) Straub, F. C. and Grabowski, H. A. 1945. Silica deposition in steam turbines. *Trans. ASME* 67:309.

(32) Herman, K. W. and Gelosa, L. R. May, 1973. Water treatment

for high pressure boilers. *Power Eng.* 77 (5):64.

(33) Amer. Soc. Testing Materials, Philadelphia, Pa. 1976 Annual Book of ASTM Standards, Part 26, D2015-66. *Standard test method for gross calorific value of solid fuel by the adiabatic bomb calorimeter.*

(34) Keenan, J. H., Keyes, F. G., Hill, P. G., and Moore, J. G. 1969. *Steam tables: thermodynamic properties of water including vapor, liquid, and solid phases.* New York: John Wiley & Sons.

(35) Glaubitz, F. Aug., 1961. Operating experience with oil-fired boilers in the combustion of sulfur-containing heating oil with the lowest possible excess air. *Mitt. Ver. Grosskesselbesitzer* 73:289. English translation in *Combustion* 34:25 (Mar., 1963).

(36) Attig, R. C. and Sedor, P. 1964. A pilot-plant investigation of factors affecting low-temperature corrosion in oil-fired boilers. *Proc. Amer. Power Conf.* 26:553.

(37) Gilbert, L. F. June 21, 1976. Precise combustion control saves fuel and power. *Chemical Engineering* 83 (13):145.

(38) Foster, R. W. Aug., 1973. Steam vs. air for sootblowing. *Combustion* 45 (2):36.

(39) Kuppusamy, N. May/June, 1978. Method of estimating sludge. *Ind. Wastes* 24 (3):32.

Chapter 7.

Complications in the Operation of Boilers

As in most industrial processes, many things can go wrong in the operation of a steam generator and its auxiliary equipment. Corrosion causes a good deal of damage, and the formation of scales and deposits on both sides of steam-generating tubes results in impaired heat transfer and eventual tube failures. In addition to paying scrupulous attention to the details of water treatment and mechanical operations, the boiler operator must also be constantly vigilant for signs of contamination in feed water or fuel. In this chapter some of the undesirable consequences of inadequate water treatment, undetected contamination, and inattention to operating procedures, are discussed suggesting, where possible, remedies or preventive measures. Procedures are also included for preparing new boilers for service, chemically cleaning dirty boilers, and reclaiming fouled ion-exchange resins.

7.1 FIRESIDE CORROSION AND DEPOSITS

Furnaces in which fuel oil is burned are subject to corrosion and the formation of deposits arising from the presence of certain chemical elements in the fuel oil. As these difficulties are not nearly so severe when coal is used for fuel, and are essentially nonexistent with natural gas, this section is confined to problems associated with burning fuel oil.

If a furnace is to be fired efficiently, it is imperative that heat and combustion losses be minimized. Avoidable losses arise from incomplete combustion of fuel, the formation of sludge in fuel oil tanks, the deposition of insulating slags on tubes and supporting hardware, and too large an excess of air to the burners. Air leaks also consume fuel unnecessarily. For instance, a 1/8-in. crack, 24 in. long in a furnace with a minus 0.3-in. draft admits 2000 ft^3/h of air. To heat this stream of air to 800 F requires some 27,000 Btu/h, which is equivalent to 26 ft^3/h of natural gas. Smoke bombs are available with which to check furnaces for leaks.

When making fireside inspections, the nature and extent of deposits should be noted in all parts of the furnace. The economizer, air preheater, and convection passages should also be inspected and any deposits found should be sampled and submitted for chemical analysis, including acidity and carbon content. Methods for doing this have been published.[1,2]

a. Residual Oils as Furnace Fuels

Many different metallic elements have been reported in residual fuel oils, the most prevalent of which are listed in Table 7.1. Iron, nickel, and vanadium usually predominate, the latter two ordinarily comprising more than 10 percent of the total ash. Vanadium, nickel, and iron may be present in oil-soluble porphyrin-metal chelated complexes and probably other complexes with nitrogen compounds and soaps. As the latter are effective emulsifying agents, they bind water or brine in extremely stable emulsions. These are especially troublesome on ships where fuel oil bunkers are customarily ballasted with seawater. It is not uncommon to find that the fuel oil from a ballasted bunker contains 0.5 percent of emulsified seawater. In addition to oil-soluble compounds in crude oil, scale, refining catalysts, corrosion products, and dirt are also sources of metallic contaminants in fuel oil.

TABLE 7.1

Metallic Contaminants in Fuel Oils

Element	Concentration range (ppm)	Form or source
Aluminum	0–25	Clay, dirt, catalyst
Calcium	0–190	Emulsified salts
Iron	0–250	Porphyrins, rust
Magnesium	0–20	Emulsified salts
Nickel	0–230	Porphyrins, corrosion products
Silicon	0–275	Clay, dirt
Sodium	5–250	Emulsified salts
Vanadium	2–300	Porphyrins

Beach and Shewmaker,[3] as a result of extraction studies, have established two classes of vanadium compounds in petroleum. Class I contains those that are extractable with aqueous pyridine; Class II contains those that are not extractable. By conducting molecular distillations with castor oil as a carrier, they estimated the equivalent atmospheric boiling points of the volatile porphyrins to be in the range of 1085 to 1200 F, with molecular weights of 543-800. From measurements of absorbency they conclude that the difference in the volatile and nonvolatile aggregates is probably in the peripheral groups rather than in the nitrogen-vanadium bonds, and that the nonvolatile porphyrins have large asphaltic or polymeric side chains that distort the nitrogen-vanadium bonds producing a noncharacteristic spectrum. Although vanadium is associated particularly with asphalt or asphalt-containing crudes, practically all crude oils contain some vanadium, its concentration ranging from a few tenths to several hundred parts per million. Nonvolatile vanadium, which concentrates in the residuum from which heavy fuel oil is produced, is extremely corrosive to refractories used in furnaces.[4] It is most destructive to fire-clay bricks; alumina and magnesia firebricks are much less affected. Vanadium oxides form low-melting eutectics with the clay, forming hard glassy slags that are very difficult to remove. Vanadium pentoxide is an active and corrosive flux, and its low melting point (1216 F), combined with its tendency to form low-melting eutectic mixtures with other compounds (notably sodium salts) contributes to the formation of slag that coats tubes, superheaters, and supporting hardware when burning fuel oils that contain vanadium.

b. Behavior of the Metallic Elements in Burning Fuel Oil

The preburner fuel oil system consists of a storage tank, an oil heater, and a strainer. Polymerization, induced by water, produces sludge in fuel oil that deposits in the oil heater and plugs the strainer and burner tips; it also represents a significant loss of usable fuel. Deposits in the burner tip interfere with atomization and can cause flame impingement on water-wall tubes. Sludge in the oil heater increases the cost of preheating and reduces the temperature of the fuel oil, which in turn increases its viscosity. In addition to inducing sludging, water causes corrosion in the storage tank and disrupts the flame at the burner tip; in extreme cases, when slugs of water reach the burner, the flame may be extinguished. Formulations are available that contain surfactant solvents to solubilize sludge, dispersants to keep it suspended, and emulsifying agents to homogenize oil and water.

Slag accumulates when metallic salts and oxides vaporized in the flame condense on cooler surfaces within the furnace, bridging tubes, blocking gas passages, and reducing heat transfer. Sodium salts lower the melting point of vanadium oxides, particles of which pass into the furnace as a sticky sublimate. Sodium chloride, sodium sulfate, nickel salts, and complex vanadates all precipitate from the vapor state, condensing as liquid droplets in the radiant and superheater sections of the furnace. Molten vanadates and particles of ash that are semimolten, or at the stage of incipient sintering, then deposit on various surfaces as a dense tenacious slag, which upon cooling is of a vitreous consistency. Other solid particles including calcium, magnesium, and iron oxides, as well as some silicates and fly ash are physically entrapped in the molten or sticky slag. Table 7.2 lists names, formulas, and fusion temperatures of several compounds that have been identified in furnace slags, all of them having melting points well below flame temperatures in the radiant section, i.e., 2600-3000 F. They condense there first, of course, because the tube temperatures are 450-650 F, then throughout the convection section as the temperature of the flue gas falls in a couple of seconds from 3000 F to about 700 F

TABLE 7.2

Chemical Compounds in Furnace Deposits

		Melting point	
Name	*Formula*	(F)	(C)
Sodium Chloride	$NaCl$	1486	808
Sodium Sulfate	Na_2SO_4	1623	884
Sodium Vanadate	Na_3VO_4	1590	866
Sodium Metavanadate	$Na_2O \cdot V_2O_5$	1165	630
Sodium Divanadate	$Na_4V_2O_7$	1209	654
Nickel Pyrovanadate	$2NiO \cdot V_2O_5$	1650	899
Vanadium Pentoxide	V_2O_5	1216	658

The results of a chemical analysis of a slag from the convection passages of a 900-psi marine boiler furnace are shown in Table 7.3.

In general, the concentration of vanadium is highest in the radiant section, sulfates are highest in the convection section, and iron is highest in

TABLE 7.3

Chemical Analysis of Furnace Slag

Component	% by weight
Vanadium Oxide, V_2O_5	44.0
Nickel Oxide, NiO	3.0
Iron Oxide, Fe_2O_3	1.0
Sodium Oxide, Na_2O	12.0
Calcium Oxide, CaO	1.0
Sulfur Trioxide, SO_3	35.0
Aluminum Oxide, Al_2O_3	0.2
Magnesium Oxide, MgO	0.4
Lead Oxide, PbO	0.4
Silicon Dioxide, SiO_2	0.6
Titanium Dioxide, TiO_2	0.3
Carbon Dioxide, CO_2	2.0
Carbon, C	0.4

the cooler sections where corrosion is increased by condensing sulfuric acid.

In addition to their insulating effect, vanadium slags attack steel, especially tube brackets, superheater hangers, and support bracelets, but much more virulent corrosion occurs in the economizer and air heater by sulfuric acid formed in the combustion of sulfur-containing fuel oils, in which the concentration of sulfur ranges from 0.2 to 6 percent. With minimum excess air, the concentration of SO_3 in flue gas seldom exceeds 10 ppm, but in the presence of contact catalysts of the first transition series (iron, nickel, vanadium, etc.), its concentration can reach 20-100 ppm.

$$S + O_2 = SO_2 \qquad (7\text{-}1)$$

$$SO_2 + \tfrac{1}{2}O_2 = SO_3 \qquad (7\text{-}2)$$

$$SO_3 + H_2O = H_2SO_4 \qquad (7\text{-}3)$$

$$H_2SO_4 + Fe = FeSO_4 + H_2 \qquad (7\text{-}4)$$

When the temperature of the exiting stack gas is below the dew point of sulfuric acid (≈ 350 F) the economizer and air heater are severely attacked by a film of concentrated sulfuric acid. Also, passages in the economizer are likely to be plugged by deposits of ferric sulfate and fly ash adhering to the wet surfaces.

c. Preburner Treatment of Fuel Oil

Salooja[5] has reviewed the additives used to inhibit the formation of slag, soot, and smoke emission, high-temperature corrosion by fused salts, and low-temperature corrosion by sulfuric acid. For many years inorganic oxides have been added to fuel oils to raise the melting points of vanadium salts and to neutralize acidity. Magnesium oxide, hydroxide, carbonate, and sulfate, and dolomitic lime are the materials that have been most widely used to induce the formation of loose powdery deposits instead of slags. These slags also convert sulfur to stable sulfates, thus reducing corrosion in cooler sections of the furnace. Dolomitic lime and other solid inorganic compounds are applied as a slurry prepared by mixing 100 lb of finely ground chemical with 180 gal of heated light fuel oil, which is then pumped continuously into the hot fuel oil line to the burners. These additives have the disadvantage of raising the ash content of the fuel oil, increasing particulate emissions from the stack, and eroding burner tips.

Lee, et al.,[6] in an elaborate investigation, found the most effective and economical ash-modifying mixture to be 10.7 percent of MgO and 10.7 percent of Al_2O_3, with particle sizes in the range of 2-7 μ, suspended in a light oil and stabilized with 0.5-2 percent of a surfactant. This combination formed a porous, friable, flaky deposit with a cubic structure softening above 2800 F. They found the most effective dosage to be 1.75 parts of $(MgO + Al_2O_3)$ to 1.0 part of vanadium, and suggest that the reaction between additive and fuel oil ash probably occurs at the tube surface rather than in the vapor phase. Ash formed this way consisted of layers of unreacted additive, additive-vanadium reaction products, and sodium vanadyl vanadate, $Na_2O \cdot V_2O_4 \cdot 5V_2O_5$, dispersed in the additive oxides. Other ash modifiers that have been proposed are zinc oxide, manganous oxide, and calcium oxide.

Combustion can be improved by adding an oxidation catalyst to the fuel oil. Manganese is quite useful for reducing smoke and soot, although it is not a replacement for balanced burners and proper fuel/air ratios. Two oil-soluble manganese compounds are favored: manganese naphthenate and methylcyclopentadienyl manganese tricarbonyl.[7] Iron additives reduce smoking in some instances; barium additives eliminate smoke completely.

Kukin[8,9] recommends 25-50 ppm of manganese in fuel oil to eliminate smoking and a combination of MgO and Mn (Mg/Mn = 6.5/1) with the ratio of magnesium to vanadium not less than 1.5. At lower ratios magnesium pyrovanadate, $2MgO \cdot V_2O_5$, can form, which is extremely corrosive to mild steel. Ramsdell, et al.[10] have described an interesting application of the mixed additive to raise superheat in furnaces that had been converted from coal to fuel oil. When this is done, a large quantity of radiant heat is absorbed by the furnace walls, which was restored by feeding magnesium oxide until the walls were insulated, then switching to manganese alone.

7.2 WATER-FORMED DEPOSITS

When a salt dissolves in water to form an ideal solution, an amount of heat is absorbed equal to the heat of fusion of the pure salt, whose solubility, therefore, increases with rising temperature.[11] Certain anhydrous salts, however, notably calcium sulfate, release heat upon dissolving, which causes their solubility to decrease with increasing temperature. Still others, such as calcium bicarbonate, are soluble in cold water, but upon heating are transformed to a slightly soluble form. Both of the salts cited are least soluble where heat flux is highest and, therefore, are prone to crystallize on surfaces where heat is transferred, a process called scaling. The resulting scale acts as insulation, preventing the transfer of heat to the water in the boiler and consequently causing the metal to overheat and eventually to fail. Scales form within a boiler when internal chemical treatment is inadequate, when high concentrations of scale-forming elements (iron, calcium, magnesium, silicates) are present in the feed water, and when burners are improperly aligned. The most insoluble depositions are often found in the superheater screen tubes, where their form and extent are greatly influenced by burner tilt.

The continuous sedimentation of suspended solids introduced as corrosion products, wear metals from bearings, impellers, and pump shafts, or mill scale and dirt left in the vessel at start-up, forms loose, soft deposits in the mud drum and water wall headers. In the absence of sludge conditioners, this type of deposit often bakes on water wall tubes, impeding the circulation of water in these critical sections. This effect is aggravated by contaminants including oil, ferric oxide, basic magnesium phosphate, and various silicates that serve as binders and nuclei for scaling at hot spots. Oil and grease are often found in accumulations from auxiliary equipment such as contact heaters, pump discharge lines, air compressors, and heat exchangers. Baked deposits in the water walls can sometimes be dislodged by hammering the tubes.

Sludge conditioners function by several different mechanisms: 1. by increasing the solubility of some slightly soluble compounds—polyacrylates are especially suited for this purpose; 2. by interfering with crystallization—polyacrylate, polymethacrylate, polymaleate, and phosphonates do this, although the latter are not particularly effective at pressures above 900 psi; 3. by limiting the size of crystals—sulfonated polymers have this effect on calcium and magnesium salts, while carboxylated polymers limit the size of ferric hydroxide particles; 4. by dispersing insoluble particles—natural dispersants include starch, lignins, alginates, and tannins. Synthetic anionic polymers such as polyacrylates, carboxymethylcellulose, and sulfonated polystyrene are useful for dispersing small crystals, somatoids, and amorphous insoluble material, all of which carry a net negative charge.

Hydraulic channeling and short retention times may cause hot process softeners to deliver make up water of high hardness. This in turn leads to plugging of filters and deposition of scale in feed lines, pumps, feed water heaters, and economizers. Also, filter alum or sodium aluminate used as coagulants in hot lime softening are soluble in the hot alkaline effluent to the extent of 3-5 ppm of aluminum. This can form analcite and other hard, dense aluminum silicate scales in boiler tubes with high heat duty. In the event of excessive hardness leakage, magnesium phosphate scale is likely to deposit in low-pressure boilers regardless of the level of alkalinity. At high pressures (600-800 psi), Holmes and Jacklin[12] report that a mixed phosphate of calcium and magnesium is formed, with some indication that the magnesium serves as a binder.

When changing internal treatments, make up water, or blowdown rate in a dirty or scaled boiler, the bond between scale and metal may be ruptured, causing scale to slough off in chips or sheets. This blocks tubes and headers, obstructs blowdown lines and other water passages, increases the concentrations of undissolved solids and alkalinity, and may initiate foaming. For this reason it is advisable to start a new treatment at concentrations well below normal, then increase the dosage gradually over a two-week period until normal residuals are reached. Dispersants may open plugged leaks in handhole gaskets and new boilers may release copious quantities of mill scale unless an alkaline boil-out has been done before starting regular treatment.

Miyakawa, et al.[13] have derived a mathematical formula for estimating the rate of deposit formation on the water side of boiler tubes. They studied the rate of deposition, the properties of deposits, and the allowable limit of thickness of deposition in 2400-psi boilers to determine when chemical cleaning should be done. In installations where this method is applicable, it is preferable to the alternatives of cutting tubes for inspection, or else chemically cleaning on a fixed schedule.

Leaking water-cooled condensers can introduce a great deal of fouling into marine and electric utility boilers. In a once-through cooling system the pressure on the water side is less than that of the atmosphere, whereas in recirculating cooling systems the water pressure is 2-3 atmospheres. Because of the pressure differential, recirculating cooling water is more likely to leak into the condensate, so it is important that tube sheets be tight and well-designed. Water-cooled heat exchangers are themselves subject to corrosion, scaling, and fouling by dirt and microbiological debris. Bacterial films increase frictional resistance to the flow of water, and extracellular polymers secreted by microorganisms adsorb suspended solids and accelerate corrosion. Recirculating cooling systems require specialized treatment with corrosion inhibitors, antiscalants, dispersants, and microbicides; this subject has been treated at length in other publications.[14,15] The more common components of water-formed deposits are listed in Table 7.4.

Although it is customary to submit samples of boiler scales and deposits to an analytical laboratory for elemental chemical analysis, Rice[16] has stressed the desirability of applying instrumental methods to identify the various phases and structures in these substances. Wet chemical analysis and emission spectroscopy determine only the elemental composition of a deposit, whereas x-ray diffraction and the polarizing microscope can be used to identify crystalline compounds and minerals. Routine x-ray diffraction measurements are used to ascertain whether a particular mineral is present, then, as a refinement, relative intensities of diffraction patterns can be estimated. To obtain quantitative results with complex mixtures, however, requires a great deal of work. Peak heights must be measured with a diffractometer, after which standards of similar composition are compared with the sample to take account of differences in the mass absorption coefficients of the various elements present. Colacito[17] has described the use of a polarizing microscope for identifying various crystallographic groups in water-formed deposits. Petrographic methods, in general, are limited to the identification of crystalline phases and are strongly dependent on the skill and experience of the microscopist.

TABLE 7.4

Components of Water-Formed Deposits

Mineral	Formula	Nature of deposit	Usual location and form
Acmite	$Na_2O \cdot Fe_2O_3 \cdot 4SiO_2$	Hard, adherent	Tube scale under hydroxyapatite or serpentine
Alpha Quartz	SiO_2	Hard, adherent	Turbine blades, mud drum, tube scale
Amphibole	$MgO \cdot SiO_2$	Adherent binder	Tube scale and sludge
Analcite	$Na_2O \cdot Al_2O_3 \cdot 4SiO_2 \cdot 2H_2O$	Hard, adherent	Tube scale under hydroxyapatite or serpentine
Anhydrite	$CaSO_4$	Hard, adherent	Tube scale, generating tubes
Aragonite	$CaCO_3$	Hard, adherent	Tube scale, feed lines, sludge
Brucite	$Mg(OH)_2$	Flocculent	Sludge in mud drum and water wall headers
Copper	Cu	Electroplated layer	Boiler tubes and turbine blades
Cuprite	Cu_2O	Adherent layer	Turbine blades, boiler deposits
Gypsum	$CaSO_4 \cdot 2H_2O$	Hard, adherent	Tube scale, generating tubes
Hematite	Fe_2O_3	Binder	Throughout boiler
Hydroxyapatite	$Ca_{10}(OH)_2(PO_4)_6$	Flocculent	Mud drum, water walls, sludge
Magnesium Phosphate	$Mg_3(PO_4)_2$	Adherent binder	Tubes, mud drum, water walls
Magnetite	Fe_3O_4	Protective film	All internal surfaces
Noselite	$3Na_2O \cdot 3Al_2O_3 \cdot 6SiO_2 \cdot Na_2SO_4$	Hard, adherent	Tube scale
Pectolite	$Na_2O \cdot 4CaO \cdot 6SiO_2 \cdot H_2O$	Hard, adherent	Tube scale
Serpentine	$3MgO \cdot 2SiO_2 \cdot H_2O$	Flocculent	Sludge
Sodalite	$3Na_2O \cdot 3Al_2O_3 \cdot 6SiO_2 \cdot 2NaCl$	Hard, adherent	Tube scale
Xonotlite	$5CaO \cdot 5SiO_2 \cdot H_2O$	Hard, adherent	Tube scale

To illustrate the possibilities of different methods of analysis, consider the elemental analysis of a thick, brown scale removed from the super-heater screen tubes in a 850-psi boiler. The results in Table 7.5 were obtained by wet chemical analysis,[2,18] the elements being reported in the conventional way, as oxides.

TABLE 7.5

Elemental Analysis of Boiler Deposit

Name	Formula	%, by weight
Silica	SiO_2	21.8
Phosphorus pentoxide	P_2O_5	23.7
Ferric oxide	Fe_2O_3	22.8
Aluminum oxide	Al_2O_3	2.4
Calcium oxide	CaO	11.8
Magnesium oxide	MgO	9.9
Sodium oxide	Na_2O	5.7

A sample of the scale was also examined by x-ray diffraction, which revealed the presence of analcite, acmite, hydroxyapatite, and hematite. In order to make some estimate of the relative proportions of these mineral structures, the values shown in Table 7.6 are first compiled.

Next a few more-or-less reasonable assumptions are made, then the probable percentages of individual compounds can be calculated by the following procedure, using the percentage compositions given in Table 7.5 and 7.6.

1. Assume all Al_2O_3 is present as analcite.

    Percent analcite $= 2.4/0.232 = 10.3$

2. Determine % Na_2O combined in analcite.

    Percent Na_2O in analcite $= 10.3 \times 0.149 = 1.5$

3. Assume remainder of Na_2O is combined in acmite.

    Percent acmite $= (5.7 - 1.5)/0.134 = 31.3$

TABLE 7.6

Percentage Composition of Minerals

	Molecular weight	% in mineral						
		Al_2O_3	Na_2O	CaO	SiO_2	Fe_2O_3	P_2O_5	MgO
Analcite	440	23.2	14.9	—	54.5	—	—	—
Acmite	462	—	13.4	—	51.9	34.6	—	—
Hydroxyapatite	1004	—	—	55.7	—	—	42.4	—
Magnesium phosphate	262	—	—	—	—	—	54.2	45.8
Ferric phosphate	151	—	—	—	—	53.0	47.0	—
Ferric oxide	160	—	—	—	—	100.0	—	—

4. Verify that all silica is accounted for.

$$\text{Percent } SiO_2 \text{ in analcite } = 10.3 \times 0.545 = 5.6$$
$$\text{Percent } SiO_2 \text{ in acmite } = 31.3 \times 0.519 = 16.2$$
$$\text{Total } = 21.8$$

5. Assume all CaO is present in hydroxyapatite.

$$\text{Percent hydroxyapatite } = 11.8/0.557 = 21.2$$

6. Assume all MgO is present in magnesium phosphate.

$$\text{Percent } Mg_3(PO_4)_2 = 9.9/0.458 = 21.6$$

7. Verify whether all P_2O_5 is accounted for.

$$\text{Percent } P_2O_5 \text{ in hydroxyapatite } = 21.2 \times 0.424 = 9.0$$
$$\text{Percent } P_2O_5 \text{ in magnesium phosphate } = 21.6 \times 0.542 = 11.7$$
$$\text{Total } = 20.7$$
$$\text{Total } P_2O_5 \text{ in deposit } = 23.7$$
$$\text{Unaccounted } = 3.0$$

8. Assume 3.0 percent P_2O_5 is present as ferric phosphate.

$$\text{Percent } FePO_4 = 3.0/0.470 = 6.4$$

9. Verify whether all Fe_2O_3 is accounted for.

$$\text{Percent } Fe_2O_3 \text{ in acmite } = 31.3 \times 0.346 = 10.8$$
$$\text{Percent } Fe_2O_3 \text{ in ferric phosphate } = 6.4 \times 0.530 = 3.4$$
$$\text{Total } = 14.2$$
$$\text{Total } Fe_2O_3 \text{ in deposit } = 22.8$$
$$\text{Difference } = 8.6$$

10. Report difference as hematite.

The final report is as shown in Table 7.7.

TABLE 7.7

Probable Composition of Boiler Deposit

Name	Formula	%, by weight
Analcite	$Na_2O \cdot Al_2O_3 \cdot 4SiO_2 \cdot 2H_2O$	10.3
Acmite	$Na_2O \cdot Fe_2O_3 \cdot 4SiO_2$	31.3
Hydroxyapatite	$Ca_{10}(OH)_2(PO_4)_6$	21.2
Magnesium phosphate	$Mg_3(PO_4)_2$	21.6
Ferric phosphate	$FePO_4$	6.4
Hematite	Fe_2O_3	8.6

7.3 SUPERHEATER AND TURBINE DEPOSITS

Circumstances that lead to foaming and carryover of boiler water result in the formation of salt deposits in superheater tubes and on turbine blades. Operational factors that can cause this sort of fouling are: too high a rate of firing, too high a steam release velocity, and sudden radical changes in steaming rate; some mechanical carryover occurs even at very low steaming rates, but it increases rapidly with steam demand. Chemical conditions that cause foaming and carryover include too high a concentration of alkalinity or total dissolved solids, contamination of boiler water by a saponifiable oil, the presence of finely divided suspended solids, notably oxides of iron and copper arising from corrosion in the condensate and preboiler systems.

In addition to operational and chemical deficiencies, there are a number of mechanical malfunctions—often overlooked—that allow boiler water to contaminate steam. The primary steam separators should be kept clean, for small deposits of rust or salts produce high velocity through the screen driers that results in carryover of boiler water. Also, separator cartridges must be bolted in tightly so there are no cracks or other openings. Handhole covers in the steam chest must be securely closed and drains must be open. The feed water inlet is usually behind the steam separation baffles with feed water being distributed over the length of the steam drum through curved tubes. If any of these inlet tubes are missing, or leak at the unions, feed water is thrown into the steam chest. Finally, care must be taken to maintain the proper water level, for if it is so low that the outlets of the steam-generating tubes in the convection section are below the surface, water is thrown across the drum into the primary steam separator. All of these maintenance items should be checked as part of the regular boiler inspection.

When boiler water passes into the superheater, the liquid immediately flashes to steam, depositing an evaporated residue of silicate, sulfate, chloride, phosphate, alkali, and whatever other salts happen to be in the water. Organic compounds, if present are carbonized on the tubes; among these are chelants, natural and synthetic polymers, and foam inhibitors. Sodium sulfite reacts in two ways, forming both hydrogen sulfide and sulfur dioxide, both of which cause corrosion in turbines.

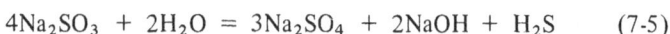

$$4Na_2SO_3 + 2H_2O = 3Na_2SO_4 + 2NaOH + H_2S \qquad (7\text{-}5)$$

$$Na_2SO_3 + H_2O = 2NaOH + SO_2 \qquad (7\text{-}6)$$

A turbine is a heat engine that converts the internal energy of steam (heat) and the potential energy attributed to its pressure to kinetic energy. Steam expands at high velocity through a fixed set of nozzles, then strikes against blades or buckets on a rotating wheel. It then expands further through a second set of nozzles, striking another stage of blades. This may be repeated many times, the nozzles and blades becoming larger as the steam pressure falls and its volume expands, until it reaches the exhaust pressure. In an impulse turbine the expansion of steam takes place in the nozzles only. In a reaction turbine there is a partial reduction in pressure in the nozzles, then a further reduction in the blades. As the velocity of the steam jet in the early stages of a turbine may reach 2000 ft/s, droplets of water or particles of chemical dust entrained in the steam are highly erosive and destructive to the nozzles and blades of the machine.

Howell and McConomy,[19] using x-ray diffraction, identified deposits of nepheline, $K_2Na_4Al_6(SiO_4)_2$, and noselite, $3Na_2O \cdot 3Al_2O_3 \cdot 6SiO_2 \cdot Na_6$-$SO_4$, in the early, high-pressure stages of a 2075-psi turbine, and α-quartz and magnetite in the low-pressure stages. Presumably, most of the elements were carried into the turbine entrained in the steam; the throttle pressure of the turbine was 2075 psi.

Silicic acid that volatilizes with steam, if in excess of 0.02 ppm SiO_2, ultimately deposits in turbines, reducing their efficiency and, in instances of severe contamination, unbalancing the rotor. Straub and Grabowski[20] found that deposits in turbines driven by steam at 600 psi were composed of sodium chloride, sodium sulfate, sodium hydroxide, and sodium silicate, and were, in general, water-soluble. At higher pressures, however, three forms of silica are deposited on blades. At the high-pressure end, where temperatures are in the range of 500–700 F, the predominant structure is sodium disilicate, $Na_6Si_2O_7$, the sodium salt of the ortho form of disilicic acid. This salt dissolves in water and is strongly hydrolyzed, so that it titrates as sodium hydroxide, although x-ray diffraction shows that the latter is not actually present in the deposit.

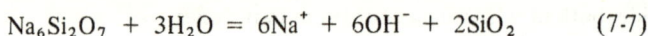

$$Na_6Si_2O_7 + 3H_2O = 6Na^+ + 6OH^- + 2SiO_2 \qquad (7\text{-}7)$$

Further along in the turbine, where the temperature falls into the range of 300–500 F, less-soluble forms of silica separate including α-quartz, cristobalite, and chalcedony. These sometimes yield to caustic washing. Finally, at the low-pressure end, where the temperature is 100–300 F, extremely insoluble amorphous silica is deposited, which usually can be removed only by mechanical cleaning, e.g., by alundum blasting. These silicates separate as they do because, although they enter the turbine dissolved in steam, as pressure and temperature fall in the course of its passage through the machine, the steam becomes supersaturated with respect to silicic acid. At the hot end there is time for crystalline structures to grow, but toward the cold end the temperature is falling so rapidly and the rate of precipitation is so fast that silica solidifies before it has time to crystallize. Amorphous forms of most chemical compounds are very slight soluble in water.

Although high steaming rates sharply increase the concentration of silica in steam, much more silica is transferred to the turbine during boiler restarts than in continuous operation. Then too, oxides formed by corrosion are at elevated concentrations in the condensate and feed water during start-ups; among these, copper is particularly troublesome when deposited on throttle valve seats or in turbines. At ultra-high pressures, sodium salts and copper oxides appear to steam-distill in a manner similar to silica, for they too pass through the superheater without any significant deposition there. During pressure reduction from 3500 to 2400 psi, however, cuprous and cupric oxides precipitate from the steam in the high-pressure stages of the turbine, seriously impairing its performance.[21]

The various types of turbine fouling discussed here, in addition to me-

chanically blocking steam passages, can also initiate stress corrosion cracking of the rotors of turbines. As these are extremely expensive to replace,
great care must be taken to see that they are supplied with clean steam.
Another related phenomenon is fatigue cracking, which occurs when metal
is repeatedly subjected to stress. Repeated cycles of stress and relief produce a pattern of striations along the surface of the fracture, which, in fer-
ritic alloys, are usually about 0.02 micron apart. In principle, by counting
the number of striations along the fracture, using an electron microscope
at 10,000 magnifications, the number of stress cycles can be determined.
If it is true that each of these lines corresponds to one stress cycle, then
the time for complete failure can be calculated by dividing the total number of cycles by the rpm of the turbine. When this is done the theoretical
time of failure turns out to be 3–4 min from the initiation of cracking. In
practice, however, the complete failure of a turbine blade takes place over
a long period of time following the initial crack formation. Fatigue cracking probably starts when the rpm of the turbine matches the resonance frequency of the blades. Because this happens infrequently, complete failure
normally takes a long time.

   As it is always worth while first to attempt to remove turbine fouling
by washing, the machines should be permanently piped to a supply of hot
treated water. The turbine to be washed must be cool before starting the
process. Water is admitted to the wash line and the drain valve is opened
to warm up the line. The throttle valve to the turbine is gradually opened
until the machine has settled at a speed of 300–350 rpm. If possible, the
turbine should be started by admitting steam through the by-pass around
the stop valve to protect against excessive speed. Finally, the valve in the
wash water line is opened to admit water until a spray of steam and water
issues from the drain in the exhaust line from the turbine. If the fouling is
susceptible to water, three hours is sufficient time to wash a turbine.

   The determination of sulfate is sometimes used to follow the washing
of turbine blades when freeing them from deposits consisting primarily of
sodium hydroxide and sodium chloride, with lesser amounts of sodium sulfate and sodium silicate. During the washing, samples are taken periodically
and sulfate is determined. Ordinarily, sodium hydroxide and sodium chloride are washed off the blades faster than sulfate, so the latter is used as an
indicator of the completeness of the washing. The concentration of sulfate
in the water used for washing must also be known. In Table 7.8 the concentrations of chloride and sulfate in samples of turbine washings taken at
30-min intervals are listed.

TABLE 7.8

Analysis of Washings from Turbine Blades

Time (min)	ppm Cl$^-$	ppm SO$_4^{--}$
0	12	5
30	235	229
60	24	71
90	16	47
120	13	20

The concentration of chloride in the wash water was 12 ppm and that of sulfate was 5 ppm.

7.4   PREBOILER CORROSION

According to Noll,[22] products of corrosion in the preboiler system are the largest single cause of tube failures in high-pressure boilers. Corrosion products include copper oxides, ferric oxide, magnetite, nickelous oxide, zinc oxide, and, occasionally, chromic oxide, all of which are introduced as particulates into feed water by the corrosion of ferrous and nonferrous alloys in the condensate and preboiler systems. In addition, x-ray diffraction has identified such complex materials as malachite, $CuCO_3 \cdot Cu(OH)_2$, ammonium carbamate, NH_2COONH_4, and hydrated basic ferric ammonium carbonate, $(NH_4)_2Fe_2(OH)_4(CO_3)_2 \cdot H_2O$, in deposits from steam lines. Condenser tubes are fabricated of admiralty brass, aluminum brass, and stainless steels. Preheaters may be made of admiralty brass, stainless steels, or in high-temperature stages, cupro-nickel, although the latter is subject to exfoliation on the steam side and to the elaboration of copper foils on the water side.[23] Economizers, shells, piping, and so on are usually of carbon steel. The approximate compositions of these alloys are as shown in Table 7.9.

Corrosion products of these alloys attach themselves to boiler tubes at places where heat transfer is high, where the flow of water is relatively slow, just beyond bends in tubes that change direction from vertical to horizontal, and at rough spots, such as butt-welded tube joints. At pres-

TABLE 7.9

Alloys Used in Preboiler Equipment

Alloy	Al	C*	Cr	Cu	Fe	Mn	Mo	Ni	Si	Sn	Zn
Admiralty brass	–	–	–	71	–	–	–	–	–	1	28
Aluminum brass	2	–	–	76	–	–	–	–	–	–	22
Carbon steel	–	0.20	–	–	99.0	0.4	–	–	0.3	–	–
Cupronickel	–	–	–	88	1.6	0.4	–	10	–	–	–
Stainless steel, 304	–	0.08	19	–	72.0	–	–	9	–	–	–
Stainless steel, 316L	–	0.03	17	–	69.0	–	2	12	–	–	–
Stainless steel, 410	–	0.15	13	–	87.0	–	–	–	–	–	–

* Maximum values

sures above 850 psi, deposition is virtually complete and these sintered deposits form a porous, insulating layer that also serves as a trap for dissolved solids, which concentrate causing pitting and caustic gouging.

Although the bulk of the oxides of iron that contaminate feed water are the result of corrosion by oxygen and carbon dioxide in condensate piping, economizers are also highly susceptible to severe pitting, as they contain very hot water. These should be inspected for evidence of corrosion by oxygen whenever the opportunity arises, for they are very expensive to replace. Copper and nickel oxides may come from the condenser or from feed water preheaters, particularly the intermediate stages. To minimize the formation of these metallic oxides, efficient operation of the deaerator verified by frequent checks for oxygen in the effluent is essential. In addition, it is necessary to maintain a continuous residual of an oxygen scavenger in the feed water, and a pH > 8 in the condensate. Corrosion in the feed water system is rapid if the efficiency of deaeration is impaired.

Corrosive attack in the deaerator is by no means unusual. It is especially likely where split-stream softening is practiced to prepare soft water with low alkalinity. In this process cation exchange using resin in both the sodium and hydrogen forms (usually with a degasifier between them) produces water that can cause corrosion in the supply line to the deaerator, particularly after the point where hot condensate is introduced. Attack can also occur in the scrubbing section, on the vent condenser tubing, and in the shell above the storage section. It is essential that all condensate pass through the spray nozzles in a deaerating heater. This is most important when deaerating condensate that has been quenched with undeaerated softened or process water that may contain substantial concentrations of dissolved oxygen.

On ships, deck lines and the drains tank after the atmospheric condenser are the major sources of iron contamination in the feed water. Steam lines that are used intermittently are subject to corrosion as is the drains tank itself. When cargo pumps are started, accumulated rust is swept by the exhaust steam through the atmospheric condenser into the drains tank, then into the feed water. Corrosion is further intensified because the de-aerator, which for reasons explained on page 179, operates satisfactorily under full load, suffers a sharp drop in efficiency at partial load, or under changing loads. Copper, nickel, and zinc are picked up primarily from the distillate condenser tubes in the evaporator and from the third and fourth stages of the feed water preheaters; the rate of copper corrosion increases with the temperature in feed water preheaters.

Iron is readily susceptible to dissolution when exposed to either oxygen or carbonic acid, but copper, nickel, and zinc are much more resistant on account of the presence of protective oxide films on their surfaces. In the presence of ammonia and oxygen, however, the protective film is destroyed with the formation of very stable ammonia complex ions. In basic ammoniacal solution the standard oxidation potentials of zinc, nickel, and copper are such that these metals are readily oxidized to their ammonia complexes by oxygen.

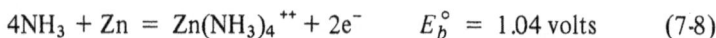

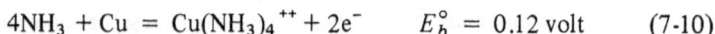

$$4NH_3 + Zn = Zn(NH_3)_4^{++} + 2e^- \qquad E_b^\circ = 1.04 \text{ volts} \qquad (7\text{-}8)$$

$$6NH_3 + Ni = Ni(NH_3)_6^{++} + 2e^- \qquad E_b^\circ = 0.47 \text{ volt} \qquad (7\text{-}9)$$

$$4NH_3 + Cu = Cu(NH_3)_4^{++} + 2e^- \qquad E_b^\circ = 0.12 \text{ volt} \qquad (7\text{-}10)$$

The rate of dissolution of copper is negligible when the pH of condensate is < 8.0 and the concentration of oxygen is < 10 ppb. Above pH 9.0, however, copper corrosion increases drastically in the presence of ammonia and oxygen. The reason for this is that as pH increases, the proportion of unionized base, i.e., free NH_3, increases in the solution. Suppose that the concentration of total ammonia is determined to be 3 ppm by the Nessler method. The percent of ionized base, i.e., NH_4^+, is given by the relation,

$$\text{Percent base ionized} = 100/[1 + \text{antilog}(pH - pK_a)] \qquad (7\text{-}11)$$

where

$$pK_a = (14.00 - pK_b)$$

For ammonia,

$$K_b = 1.77 \times 10^{-5} \qquad \text{(at 25 C)}$$
$$\log K_b = \bar{5}.25$$
$$pK_b = 4.75$$
$$pK_a = (14.00 - 4.75)$$
$$= 9.25$$

Thus, at pH 8.0,

$$\text{Percent base ionized} = 100/[1 + \text{antilog}(8.00 - 9.25)]$$
$$= 94.7$$

From which

$$\text{Percent free NH}_3 = (100 - 94.7)$$
$$= 5.3$$

Or, in ppm,

$$3 \times 0.053 = 0.16 \text{ ppm NH}_3$$

At pH 9.0,

$$\text{Percent base ionized} = 100/[1 + \text{antilog}(9.00 - 9.25)]$$
$$= 64.0$$

And,

$$\text{Percent free NH}_3 = (100 - 64)$$
$$= 36$$

In ppm,

$$3 \times 0.36 = 1.1 \text{ ppm NH}_3$$

Thus, there is approximately seven times the concentration of free NH_3 at pH 9 that there is at pH 8. In accordance with the law of mass action, half-reactions (7-8), (7-9), and (7-10), therefore, have the greater tendency to proceed as written, the higher the pH. For some purposes, the results calculated from Eq. (7-11) must be corrected for the effects of ionic strength; this has been discussed in an earlier publication.[14]

Ammonia is introduced into condensates as a decomposition product of hydrazine, and to a lesser extent, of morpholine and cyclohexylamine. The thermal decomposition of hydrazine begins at 400 F; it is about 60 percent complete at 550 F. Several equations can be written for the decomposition, but the most likely reaction yields one mole of ammonia per mole of hydrazine.

$$2N_2H_4 = N_2 + H_2 + 2NH_3 \qquad (5\text{-}17)$$

Several actions can be taken to limit the formation of corrosion products and to nullify their ill effects in boilers. Preboiler corrosion can be curtailed by removing oxygen, by using filming amines or neutralizing amines, by selecting appropriate materials of construction, and by filtering the condensate. The most important of these steps, as already mentioned, is the removal of oxygen by deaeration and by adding an oxygen scavenger—hydrazine in boilers operated at 900 psi and above. Hydrazine should be added to the storage section of the deaerator at a concentration 1.5 times the concentration of oxygen in the deaerator effluent. In instances of severe copper corrosion in the condenser, hydrazine can be injected into the steam to the low-pressure turbine, as is done on ships. As hydrazine does not survive passage through the superheater, no protection against oxygen is available in and beyond the condenser, unless a supplemental amount is added at the low-pressure turbine. Only 15-20 percent of this additional hydrazine is lost in its subsequent passage through the deaerator.

Filming amines, under the best of conditions, provide only about 75 percent protection in condensate piping, and they have the further disadvantage of forming black, waxy, magnetic deposits that plug strainers and steam traps, cause overspeed trips to stick, and accumulate within the boiler, sometimes impeding circulation through water passages. Morpholine, cyclohexylamine, or a combination of the two are used to elevate the pH of condensate to minimize corrosive attack by carbonic acid. These neutralizing amines can be conveniently added, mixed with hydrazine, to the storage section of the deaerator, or alternatively, to the deaerator outlet. They should be added in an amount sufficient to give a pH of 8.5-8.8, measured at the discharge of the condensate pump.

Another possible source of particulate Fe_3O_4 in high-pressure boilers arises from the thermal transformation of magnetite to wustite (FeO). A film of magnetite forms on steel in contact with water at temperatures > 250 C (482 F).

$$3Fe + 4H_2O = Fe_3O_4 + 4H_2 \qquad (2\text{-}10)$$

The partial-pressure diagram for iron-oxygen given by Garrels and Christ[24] shows that above 570 C (1058 F), Fe_3O_4 cannot exist, but is transformed to FeO. Thus, if steel covered by a film of magnetite is heated above this temperature, a layer of FeO forms between the steel and the magnetite and the latter may chip off. This reaction is a matter of interest rather than of practical importance, as steel could not long survive this extreme temperature.

Neat[23] and Long[25] have both recommended the replacement of non-ferrous alloys in the preboiler system by constructing surface condensers of stainless steel and extraction feed water heaters of carbon steel, which would eliminate copper and zinc from the feed water. Stainless steel, however, is especially susceptible to pitting under deposits of any sort and is also subject to stress-corrosion cracking in the presence of chloride ion; this latter susceptibility can be reduced by including molybdenum in the alloy, as in 316L stainless steel. Filtering particulate matter from condensate has already been discussed in Chapter 6.

7.5 INTERNAL CORROSION AND TUBE FAILURES

In an industry-wide survey, Klein, et al.[26] found that the accumulation of metal oxides, discussed in the previous section, accounted for about 50 percent of the corrosion in high-pressure boilers. Condenser leaks or contamination by acids produced 30 percent, and faulty design contributed to 20 percent of the instances of corrosion reported. As the majority of high-pressure boilers are under either coordinated phosphate-pH or volatile treatments, most of the corrosive attacks reported were embrittlement failures caused by hydrogen or hydrogen ion, rather than ductile gouging by hydroxyl ion. In the electric utility industry, peaking steam generators, which are operated intermittently to meet high demand, collect a great deal more metal oxides than do base-loaded boilers, which operate continuously at constant load.

In general, boilers are constructed of carbon steel, a two-phase mixture of ferrite and pearlite.[27] Ferrite is a solid solution of a small concentration of carbon (≈ 0.2 percent) in α-iron; pearlite is a mixture of ferrite and cementite, Fe_3C, in a laminar arrangement. The structure of carbon steel can be altered, weakened, or destroyed by chemical reactions and also by heat. Oxygen introduced in the feed water produces random pitting in economizers, downcomers, and at the water line in steam drums. At temperatures above 1300 F, oxygen weakens steel by intergranular oxidation along grain boundaries, and by a process called decarburization, in which carbon in the ferrite phase is oxidized to carbon dioxide. At lower temperatures, oxygen first converts the protective film of magnetite to ferric oxide, then oxidizes the exposed steel to γ-FeOOH.[28]

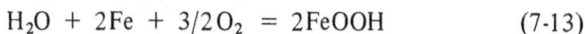

$$2Fe_3O_4 + 1/2O_2 = 3Fe_2O_3 \qquad (7\text{-}12)$$

$$H_2O + 2Fe + 3/2O_2 = 2FeOOH \qquad (7\text{-}13)$$

The gelatinous γ-FeOOH formed initially is soon dehydrated to Fe_2O_3, which provides no protection for the metal surface.

$$2FeOOH = Fe_2O_3 + H_2O \qquad (7\text{-}14)$$

Special attention must be given to protecting a boiler from oxygen during standby and layup.

Severe damage can result from the concentration of boiler salines under porous deposits of iron and copper oxides introduced as products of pre-boiler corrosion. These deposits act as diffusion barriers that facilitate the concentration beneath them of otherwise-soluble salts, notably alkalis. Concentrated hydroxyl ion dissolves the protective film of magnetite, forming ferrite and hypoferrite ions.

$$Fe_3O_4 + 4OH^- = 2FeO_2^- + FeO_2^{--} + 2H_2O \qquad (2\text{-}14)$$

This phenomenon, called caustic or ductile gouging, often takes place under deposits where heat transfer rates are high, although Noll[29] points out that a deposit need not always be present for this type of attack to occur. Gouging is especially common in boilers under the caustic reserve method of alkalinity control. A special form of corrosion induced by hydroxyl ion, stress-corrosion cracking, or caustic embrittlement, has been discussed in detail in Section 2.3. It is characterized by continuous cracking originating at a surface and continuing along the grain boundaries in steel. This phenomenon is common in riveted boiler drums in which stress is introduced by driving the rivets, and in which leakage of boiler water can occur, resulting in the local concentration of boiler salines. It is uncommon in welded stress-relieved boilers.

The most critical of the preboiler contaminants is copper. If the element enters the boiler as the cupric ammonia complex ion, $Cu(NH_3)_4{}^{++}$, it attacks steel directly by electrolytic reduction.

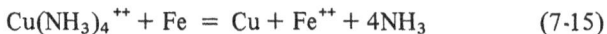

$$Cu(NH_3)_4{}^{++} + Fe = Cu + Fe^{++} + 4NH_3 \qquad (7\text{-}15)$$

The plated copper then becomes a cathode to which the surrounding iron behaves as a sacrificial anode, leading to extensive corrosion and thinning under and around the layer of copper metal. The accumulation of boiler and preboiler corrosion products can be eliminated by preoperational boil-out, by removing the contaminants from the feed water, and, if necessary, by chemical cleaning.

Contaminants that lower the pH of boiler water can cause rapid and severe corrosion. Acid left under resistant deposits after chemical cleaning, for instance, becomes extremely virulent upon heating.

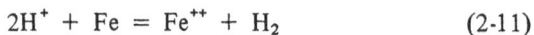

$$2H^+ + Fe = Fe^{++} + H_2 \qquad (2\text{-}11)$$

In many cooling waters used for condensing steam, the magnesium hardness exceeds the bicarbonate alkalinity. In the event of a condenser leak, when this water enters the boiler, the pH may be lowered sufficiently to initiate acidic attack, particularly in boilers on volatile treatment.

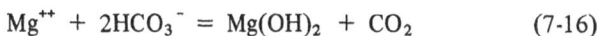

$$Mg^{++} + 2HCO_3{}^- = Mg(OH)_2 + CO_2 \qquad (7\text{-}16)$$

Acid attack, mainly transgranular, is especially intense on stressed metal—field welds, for example; destruction is intensified because stressed metal is anodic to unstressed metal. Over-treatment with the chelants, ethylenediaminetetraacetate and nitrilotriacetate, causes thinning at the water line in the steam drum and in places where water velocity is high; the corroded surface is characteristically smooth and darkened.

In areas such as horizontal runs and roof tubes, where the metal is partly dry or steam-blanketed, the protective film of magnetite can react directly with steam, which then attacks the underlying steel.

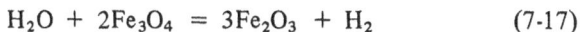

$$H_2O + 2Fe_3O_4 = 3Fe_2O_3 + H_2 \qquad (7\text{-}17)$$

$$2Fe + 3H_2O = Fe_2O_3 + 3H_2 \qquad (7\text{-}18)$$

In addition to wastage of metal by the reaction seen in Eq. (7-18), the evolved hydrogen itself is harmful. Hydrogen embrittlement, in which discontinuous intergranular fissures are formed in steel as a result of internal pressure generated by a mixture of hydrocarbon gases, is a secondary form of corrosion caused by hydrogen gas released in a primary corrosion by steam, hydroxyl ion, or hydrogen ion. At relatively low temperature, gaseous hydrogen diffuses into steel and reacts with carbon to form various hydrocarbons of low molecular weight. A hard, brittle veneer is always evident on steel damaged by hydrogen. This barrier prevents the escape of the gas from the surface of the metal. Ames and Lux[30] report that hydrogen damage arising from contamination is infrequent in boilers treated with phosphate; they also state that the presence of hydrogen gas in steam is not a reliable indicator of hydrogen cracking.

Van Brunt[31] has listed fifteen types of tube failure, several of which result from corrosion mechanisms mentioned above. In addition to these, there are several changes in crystalline structure that occur in overheated steel that can lead to failure. Tubes that are continuously subjected to temperatures in the range of 850-950 F undergo spheroidization, in which the pearlite phase disappears as the laminar cementite is gradually transformed to spherical grains. Goodstine and Kurpan[32] mention that hydrogen damage is common under dense deposits where brittle fracture may occur with an entire section blowing out of a tube. Tubes also are frequently destroyed by flame impingement, as a result of improperly aligned burners, or by a restricted flow of water; both of these cause rapid overheating and consequent failure of the metal. Another source of tube failure is crevice corrosion under backing rings. When butt-welding two sections of tubing it is common practice to insert a steel ring in the ends of the tubes to be welded to align the two sections. More often than not, an open space is left between the backing ring and the internal surface of the tube, where boiler water can be trapped and concentrated. As the backing ring often introduces strain, stress-corrosion cracking may ensue. This source of trouble can be eliminated by using chill rings, which are usually made of copper to conduct heat away, and which are removed when the weld is completed. Similar corrosion has been seen in steam lines following desuperheaters to which demineralized water is fed. Sodium leakage from the cation exchanger introduces small amounts of sodium hydroxide into the atemperating water that ultimately concentrates in crevices, initiating caustic embrittlement. Caustic also attacks handhole gaskets and seats if covers are not seated accurately and tightly secured; as mentioned before, phosphate treatments are very effective in sealing the resulting leaks.

Noll[22] points out that many of the causes of tube failures can be ascertained by metallographic examination. These include decarburation, hydrogen embrittlement, spheroidization, intergranular oxidation, stress-corrosion cracking, and caustic gouging. This sort of examination is not worthwhile for tubes containing scales, baked deposits, or other indications of overheating where the reason for failure is obvious.

7.6 CONTAMINANTS IN BOILER FEED WATER

In eariler sections the effects of acids, condenser leaks, and preboiler corrosion products in the boiler and turbine that were introduced in the feed water were discussed. It is highly desirable to have an immediate warning when the feed water becomes contaminated, so that remedial action can be taken; e.g., adding trisodium phosphate to the boiler when acid contamination is present. Conductivity alarms answer this purpose well in most instances, although contamination of feed water by oil not only does not affect conductivity, it inactivates the sensing cell by coating the platinized electrodes. Miscellaneous hydrocarbons, grease, and other oils enter the condensate through leakage in steam-heated fuel oil heaters, or in heat exchangers used in various processes; cylinder oil from steam-driven machinery is also common in the condensate.

Oil contamination in a boiler produces the following adverse effects:
1. Contributes to film boiling by making steam-generating surfaces nonwettable.
2. Serves as a binder of insoluble particles leading to scaling in hotter sections of a boiler.
3. May deposit coke on boiler tubes at places where heat flux is especially high.
4. Interferes with the action of natural sludge conditioners (lignins, tannins, quebracho, etc.) although the polyacrylates function well in the presence of oil.
5. Causes foaming that is particularly serious if the oil is saponifiable, for instance, a vegetable oil that forms a soap in hot alkaline solution.
6. Collects boiler sludge into spherical masses ("oil balls") that block headers or bake on water wall tubes.

Methods for removing oil are described in Section 3.3. Precoat filters provide the most efficient removal of oil if the alumina gel is prepared by mixing sodium aluminate and filter alum.

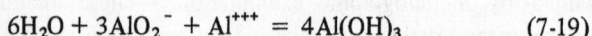

$$6H_2O + 3AlO_2^- + Al^{+++} = 4Al(OH)_3 \qquad (7\text{-}19)$$

Some organic chemicals, such as phenols, glycols, and higher alcohols cause objectionable odors to issue from deaerator vents, steam ejectors, or steam traps. Others lay down varnish in high-pressure turbines and sticky films in the low-pressure stages. If a condensate is found to be or suspected to be contaminated with organic material it should be dumped to the sewer and the boiler should be blown down. Oil and other organic material can be removed from a badly fouled boiler by the preoperational cleaning procedure specified in the next section.

The effect of large amounts of ammonia accidentally introduced into a boiler is of interest. This can happen either through failure of injection equipment or through contaminated condensate returns from certain manufacturing processes. The following reactions can occur.

$$2NH_3 + CO_3^{--} + H_2O = (NH_4)_2CO_3 + 2OH^- \qquad (7\text{-}20)$$

$$3NH_4^+ + OH^- + CO_3^{--} = (NH_4)_2CO_3 + NH_4OH \qquad (7\text{-}21)$$

In both cases this is followed by

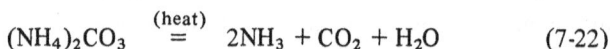

$$(NH_4)_2CO_3 \overset{\text{(heat)}}{=} 2NH_3 + CO_2 + H_2O \qquad (7\text{-}22)$$

It is seen from these reactions that with an excess of ammonia, the carbonate alkalinity is reduced to zero, while with ammonium salts the total alkalinity is reduced to zero. These equations are only formal; actually ammonium carbamate is probably also involved in the mechanism of the decomposition.

7.7   CHEMICAL CLEANING PROCEDURES

Now several chemical cleaning methods applicable to a steam generating plant are considered. These procedures are much more rapid and economical than mechanical cleaning by turbining, hammering, and brushing, all of which require a great deal of labor. Among the topics covered in this section are preoperational cleaning, in-service cleaning, and acid cleaning of boilers, as well as reconditioning of fouled ion-exchange resins.

a. Preoperational Cleaning of Boilers

New boilers and other equipment used for exchanging heat must be subjected to an alkaline "boil-out" before being put in service, in order to remove cutting oils, mill scale, grease, lubricating oils, pipe-threading compound, and drawing compounds introduced during manufacture and installation of the equipment. The purpose of this is to expose the surface of the metal so that a normal uniform protective film of magnetite can form when the unit is put in service. If preoperational cleaning is neglected, pitting and generalized corrosion occurs because covered areas of metal are anodic to clean areas.

Suppose, for instance, it is necessary to condition a newly installed boiler having a capacity of 25,000 gal of water. A 50 percent solution of caustic soda along with commercial soda ash is used to prepare the alkaline cleaning solution, which is pumped into the boiler and diluted to 25,000 gal with treated make up water. The final concentration of each chemical should be about 3000 ppm. The number of pounds required of each chemical is

$$\text{lb chemical} = (3000 \times 25{,}000 \times 8.34)/10^6$$

$$= 625.5$$

As 50 percent caustic soda contains 6.365 lb NaOH/gal, the volume needed is

$$625.5/6.365 \cong 98 \text{ gal}$$

The solubility in water of soda ash is 14 percent by weight, at 1.146 specific gravity.[33] Thus, a saturated solution contains

$$(14.0 \times 1.146 \times 8.34)/100 = 1.338 \text{ lb } Na_2CO_3/\text{gal}$$

The required amount of soda ash, 625.5 lb, can therefore be dissolved in a 500-gal mixing tank and then pumped into the boiler. In practice, the boiler should be half-filled with treated make up water and the prepared solution of soda ash added, after which the concentrated caustic soda is injected and the boiler is filled with water. The pressure is raised and the solution is circulated for 24 h, during which time the concentrations of NaOH and Na_2CO_3 are occasionally checked by the procedure in Chapter 8, page 265; more chemicals are added, as required, to keep the concentrations of both at 3000 ppm. After dumping the alkaline solution and rinsing once or twice with deaerated fresh water, the unit is ready for service.

In the United States, it is customary to follow the alkaline wash with acid cleaning, particularly when commissioning boilers to be operated in excess of 2000 psi. Associated equipment including deaerator, condenser, and stage heaters may also be acid-cleaned. Because of the presence of stainless steels, citric or hydroxyacetic and formic acids are used instead of hydrochloric acid.

b. In-Service Boiler Cleaning

Experienced boiler operators are aware that relatively minor changes in the composition of make up water, alterations in internal treatment, or radical changes in blowdown rate can cause scale to slough or spall from a scaled boiler, blocking tubes and headers, plugging blowdown lines, and obstructing circulation. For this reason, it is advisable to be conservative when changing from one internal treatment to another. A related concept is that of removing scales by deliberately altering chemical treatment, a process that has become known as in-service cleaning. It is not always possible to predict the effect of a chemical on a scale, but, in general, phosphate attacks calcium carbonate rapidly, causing foaming (effervescence, actually, as much carbon dioxide is released) and high alkalinity.

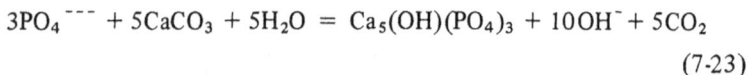

$$3PO_4^{---} + 5CaCO_3 + 5H_2O = Ca_5(OH)(PO_4)_3 + 10OH^- + 5CO_2 \tag{7-23}$$

In the presence of carbonate, phosphate also transforms calcium sulfate very rapidly.

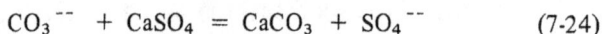

$$CO_3^{--} + CaSO_4 = CaCO_3 + SO_4^{--} \tag{7-24}$$

Eq. (7-24) is followed by the reaction shown in Eq. (7-23). If the intention is to remove calcium sulfate scale alone, however, it is preferable to use sodium hydroxide, as it is then easier to control alkalinity and the foaming caused by the evolution of carbon dioxide is avoided.

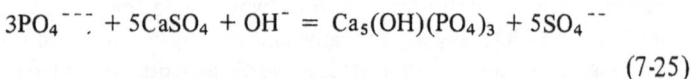

$$3PO_4^{---} + 5CaSO_4 + OH^- = Ca_5(OH)(PO_4)_3 + 5SO_4^{--} \tag{7-25}$$

Calcium silicate is slowly dissolved by phosphate, but elevated alkalinity more rapidly attacks this and other silicates; the addition of silicates themselves is effective for fluidizing magnesium sludges.

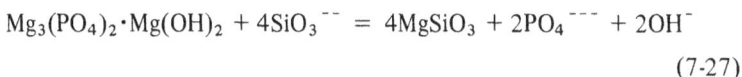

$$CaSiO_3 + 2OH^- = Ca^{++} + SiO_4^{-4} + H_2O \qquad (7\text{-}26)$$

$$Mg_3(PO_4)_2 \cdot Mg(OH)_2 + 4SiO_3^{--} = 4MgSiO_3 + 2PO_4^{---} + 2OH^-$$
$$(7\text{-}27)$$

It is unlikely that any of these reactions go to completion in a boiler; their major effect is to loosen the connection between scale and metal. When contemplating a change in treatment in an unusually dirty boiler containing voluminous deposits, extensive scale, or copious amounts of iron oxides and copper, it is advisable to acid-clean, as described further along in this section.

Herman and Gelosa[34] have alluded to an internal treatment for sludge removal wherein polymers, organics, sequestrants, and iron sequestrant-dispersant are added at unspecified concentrations to accomplish a slow, on-stream cleanup. To be a little more specific as to the details of this scheme: lignins or carboxymethylcellulose are used in their usual capacity as protective colloids to disperse sludge; to supplement the dispersants, a few parts per million of chelant are included, which seems to assist in loosening existing deposits; hydroxyethylidenediphosphonate, or other phosphonates, may be included to assist in disintegrating deposits coagulated by iron. Approximately three months are required to dissipate sludges and deposits that are at all susceptible to the treatment.

Edwards and Rozas[35] have reported the pioneering work of the Dow Chemical Company, starting in 1959, in continuous internal treatment of boilers with EDTA. Subsequently,[36] the possibilities of this treatment for in-service cleaning became apparent. Although chelants cannot sequester iron oxides, they do chelate calcium and magnesium in deposits and sludge in which iron functions as a binder. This releases particles of iron oxide, which are then removed in the blowdown. Thus, within a few hours of initiating a chelant treatment in a dirty boiler, quantities of suspended iron oxide are seen in the blowdown water. Also, when a phosphate program is replaced by chelant, the concentration of phosphate ion in the boiler is likely to be higher for several days than it was before discontinuing the addition of phosphate, because of some dissolution of calcium and magnesium phosphates. When using chelants for in-service cleaning, the boiler operator should observe the precautions discussed in Section 5.2b, which apply to any chelant treatment, however described by the vendor.

c. Special Cleaning Methods

The periodic cleaning of boilers with acids and other chemicals is now an accepted routine procedure in the United States, although many engineers, especially in Europe, recommend against it. In Germany, for instance, the acid cleaning of a steam generator is a decidedly rare event, whereas in the United States many boilers are cleaned every two or three years. Deposits in moderate-pressure boilers usually are composed of hematite, magnetite, silicates, and alkaline earth salts; at pressures above 2000 psi laminar scales of hematite, cuprous oxide, and metallic copper are common. The location and thickness of copper scales can be determined by x-ray radiography, but when other deposits are present, Wackenhuth and Richards[37] recommend removing a water wall tube to determine whether chemical cleaning is necessary and, if so, to find an effective solvent for the purpose. The choice of cleaning agent is to some extent empirical, so some sort of performance tests in a laboratory may be desirable to assess solubility in various chemicals, to find the most suitable temperature, and to determine the efficiency to be expected.[38]

The solvent most frequently chosen to clean boilers is a dilute solution of hydrochloric acid to which an inhibitor is added to minimize attack on the metal being cleaned. Hydrochloric acid is relatively inexpensive and is especially effective for dissolving ferric oxide because the driving force of the neutralizing reaction

$$Fe_2O_3 + 6H^+ = Fe^{+++} + 3H_2O \qquad (7\text{-}28)$$

obtained with any acid is augmented by the formation of the very stable hexachloroferric ion:

$$Fe_2O_3 + 6H^+ + 6Cl^- = FeCl_6^{---} + 3H_2O \qquad (7\text{-}29)$$

The effects of both hydrogen and chloride ions are thus brought to bear in dissolving ferric oxide in hydrochloric acid.

The reaction

$$2H^+ + Fe = Fe^{++} + H_2 \qquad (7\text{-}30)$$

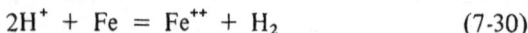

is opposed by an irreversible potential called the hydrogen overvoltage, which is markedly affected by the condition of the metallic surface. In addition, the corrosion rate of a metal in acidic solution is governed by the rate of reduction of hydrogen ions to hydrogen gas at cathodic sites. This too is influenced by surface properties and by mechanical barriers that interfere with the adsorption of hydrogen atoms, the first step in the formation of hydrogen gas. Nitrogen bases (triamylamine, pyridine, quinoline, acridine, imidazoline, and many others) inhibit the attack of hydrogen ions on steel by raising the hydrogen overvoltage on cathodic surfaces. Inhibition by nitrogen bases is enhanced when they are combined with acetylenic alcohols, such as propargyl alcohol (propyne-1-ol-3), $HC{\equiv}CCH_2OH$, and 2-butyne-1,4-diol, $CH_2OH{-}C{\equiv}C{-}CH_2OH$, which are adsorbed as a film on metallic surfaces. Another combination claimed to be useful[39] is thiourea with one of the Mannich bases.[40] The latter are synthesized by condensing a carbonyl compound having an acidic hydrogen atom in the α-position with formaldehyde and a primary or secondary amine. For example, acetone condenses with formaldehyde and dimethylamine to form the Mannich base ω-dimethylaminobutanone.

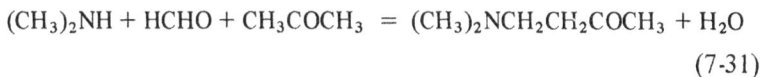

$$(CH_3)_2NH + HCHO + CH_3COCH_3 = (CH_3)_2NCH_2CH_2COCH_3 + H_2O$$

$$(7\text{-}31)$$

Acid cleaning may be undertaken for any of several reasons: 1. to reduce the thickness of the magnetite film and to dissolve deposits of copper, zinc, and iron oxides; 2. to prepare the boiler for a change in internal treatment; 3. to achieve an immediate improvement in heat transfer; 4. to avoid troublesome unscheduled shutdowns. Rice[41] has reported, however, that repeated acid cleanings can damage areas with heavy local stress such as field welds, improperly rolled tubes, cold-worked metal, bolt threads, and draw marks on tubes. Longitudinal cracking, roughening of tube ends, deep etching, pitting, and honeycombing have all been seen after acid cleaning, the severity of attack being influenced to a great extent by the kind of deposit in contact with the boiler metal when cleaning commences. Another complication arises when deposits are incompletely removed and some of the acid cleaning solution is trapped beneath them. When the boiler is subsequently fired, serious hydrogen damage occurs in these locations. For these reasons, it is advisable to check the soundness of tubes and the extent of corrosion by ultrasonic testing following an acid cleaning.

Boilers operated at low to moderate pressures are likely to contain phosphate sludges, alkaline earth salts, thickened layers of Fe_3O_4, and silicate scales. Most of these dissolve readily in dilute inhibited hydrochloric acid, but if silicates are present, they are not appreciably affected unless ammonium bifluoride is added to the acid solution. A satisfactory cleaning solution for boilers that do not contain deposits of copper or copper oxides is prepared as follows:

> 5.0-7.5 percent HCl*
> 0.5-1.5 percent NH_4HF_2
> 0.2-0.3 percent proprietary inhibitor
> 0.03 percent nonionic wetting agent, e.g.,
> an alkylarylpolyethoxy alcohol

The corrosivity to carbon steel of the stronger of the above mixtures was found to be 260-350 mpy (0.03-0.04 mph), without any pitting.

The strengths of hydrochloric acid and ammonium bifluoride to be used depend upon the history of the boiler to be cleaned. In general, if a boiler has not been cleaned for several years, or if inspection reveals extensive deposits, a solution containing 7.5 percent of hydrochloric acid and 1.5 percent of ammonium bifluoride is appropriate. On the other hand, if the boiler has been cleaned within the preceding two years, if the feed water is known to be satisfactory and no upsets have occurred, and if only light deposits are found on inspection, then the lesser strengths of 5.0 percent hydrochloric acid and 0.5 percent ammonium bifluoride are sufficient. Atwood[38] has shown that hydrochloric acid is the only cleaner that dissolves scale in a reasonable length of time under the static condition that exists when a boiler is filled and soaked without circulation; all other common cleaning agents must be circulated at at least 1 ft/s to remove scale completely at a reasonable rate.

If organic sludge conditioners such as lignins, tannins, or quebracho have been used for internal treatment, or if the presence of soil is suspected, an alkaline boil-out should be done before attempting to acid clean. This is accomplished by operating the boiler at about 100 psi while injecting the alkaline solutions described in Section 7.7a, through the chemical injection pump(s), then blowing down the boiler until the pH of the water is reduced to about 10. Some engineers recommend that all new boilers also be acid-cleaned after alkaline boil-out.

Preparatory to acid-cleaning, a boiler is brought down and the tubes are washed with hot boiler feed water to remove loose debris. When the wash is completed, vents are opened, draft fans are started, and the boiler is

---

* Some boiler manufacturers limit the concentration of hydrochloric acid to 5 percent.

cooled to 160–180 F, an attempt being made to achieve this temperature range by the time the cleaning contractor is scheduled to arrive. All furnace openings are closed, including drafts, dampers, and burner doors, to retain heat in the boiler; the latter should never be preheated by direct firing.

Boilermakers next isolate all lines and equipment that should not be exposed to hydrochloric acid (stainless steel, in particular) then remove the blowdown valves on the water wall headers, replacing them with 2-in. screwed connections. A drain is installed on the lower drum and a 2-in. screwed connection is also installed either on the steam drum just above the normal water level, or else on the upper arm of the level controller. After repairing any leaks that may have developed, the boiler is turned over to the chemical cleaning contractor. The contractor connects hoses from the discharge manifold of his chemical solution pump to each of the blowdown nozzles provided; another hose connects the steam drum to the solution tank from which the chemical pump takes suction.

Sometime prior to filling the boiler with inhibited acid solution, the superheater is back-flushed with water until it is filled and overflows into the steam drum. This is done to prevent the inadvertent entry of acid should the steam drum be overfilled and to prevent exposure to acidic vapor should the temperature of the boiler be so high that the acid solution boils. Neither eventuality is likely if an experienced contractor is employed, but the precaution is worthwhile.

Before pumping the cleaning solution into the boiler, it is the responsibility of the contractor to verify that the temperature of the boiler is in the specified range. This is usually done by measuring the temperatures at the ends of the boiler drums, although this is inexact, as the ends of the drums are the thinnest accessible parts of the boiler, and thus do not reflect what the temperature might be in the interior, where there is a great mass of metal not directly exposed to the cooling fans. Because of this, the contractor may first fill the boiler with tempering water heated to a few degrees below the desired temperature if the readings at the drum ends are too high, or a few degrees above the desired temperature if the readings are too low. If, after passing through the boiler, the temperature of the tempering water is in the correct range, the boiler may be filled with acid solution, and pumped in at a little above or below the desired temperature, according to circumstances.

Because of the large volume of cleaning solution required, a flow of water is started into the boiler and 37 percent hydrochloric acid, suitably inhibited, is blended into the water to give the desired concentration. This requires frequent titrations of acid strength and adjustments of the flow of acid during the filling. The boiler is filled through the lower blowdown connections until the solution overflows into the tank truck through the hose connected to the steam drum. The usual procedure is to allow the boiler to soak without circulation for about 6 h, which is sufficient time for hydrochloric acid to dissolve most water-formed deposits. It is better practice, however, to use the filling pump to maintain a slow circulation throughout the 6-h period. The pump augments thermal circulation, continuously delivering fresh acid to the internal surfaces. Even though the total volume is much greater than the capacity of the inlet nozzles, their through-put is still sufficient to provide slow circulation and also to reduce the temperature of any hot spots that might exist. In some instances, where scale is exceptionally heavy, a longer treatment might be needed, but this can be determined only if the boiler has first been opened for inspection. As circulation is slow and the volume of acid is large compared to the weight of scale contacted, there is no need to measure the concentration of acid during the soaking period.

When cleaning is complete, the acid solution is forced out through the lower drain hose, under 5–10 psi of nitrogen, into a vacuum truck for disposal. After the acid has been drained, the boiler is filled with hot deaerated water, drained, filled, and drained again, in each instance under a nitrogen blanket to prevent flash rusting. It is essential to keep the inside of the boiler as free from air as possible to avoid extensive oxygen pitting along mandrel or draw markings. To the unaided eye these appear to be cracks, but when magnified, they are seen to be strings of pits. Greenberg[42] recommends a rinse with 0.2 percent hydrochloric acid (pH $\approx$ 2) to dissolve hydrous ferric oxide, which precipitates by hydrolysis at pH 3, or above.

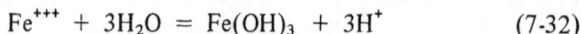

$$Fe^{+++} + 3H_2O = Fe(OH)_3 + 3H^+ \qquad (7\text{-}32)$$

Finally, a neutralizing/passivating solution is pumped into the boiler and allowed to soak for an hour. Alternatively, the boiler is fired lightly to about 100 psi, then cooled and drained. This treatment neutralizes residual acidity and provides a passive surface on the cleaned metal that resists rusting when the boiler is opened for inspection and repairs. A suitable passivating solution contains the following ingredients:

$$0.5 \quad \text{percent } Na_2CO_3$$

0.25 percent NaH_2PO_4
0.25 percent Na_2HPO_4
0.5 percent $NaNO_2$

Note that if this solution is exposed to an excess of acid, the very toxic gas, nitrogen dioxide, is evolved, as evidenced by brown fumes of NO_2. Even at low concentrations this gas rapidly causes severe damage to the lungs.

$$2H^+ + 2NO_2^- = NO + NO_2 + H_2O \qquad (7\text{-}33)$$

As scales in high-pressure boilers are, in the main, products of preboiler corrosion, they typically contain oxides of iron, copper, zinc, nickel, and sometimes aluminum. Metallic copper is often present and spinels have also been identified by x-ray diffraction.[37] The latter are very hard aluminates or ferrates of magnesium, zinc, copper, or nickel, e.g., $CuAl_2O_4$. These deposits are usually found on the inside surfaces of boiler tubes on the sides exposed to the burners. Special cleaning procedures, often in two steps, are necessary for boilers containing deposits of curpous oxide, cupric oxide, and metallic copper. When hydrochloric acid is added to cupric oxide, the latter dissolves, then cupric ion is immediately reduced to copper on steel surfaces, solubilizing an equivalent amount of iron.

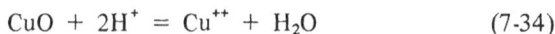

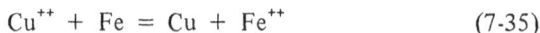

$$CuO + 2H^+ = Cu^{++} + H_2O \qquad (7\text{-}34)$$

$$Cu^{++} + Fe = Cu + Fe^{++} \qquad (7\text{-}35)$$

A method has been disclosed[43,44] for blocking the reaction in Eq. (7-35) by adding thiourea to the hydrochloric acid cleaning solution. The organic compound reduces cupric ion to cuprous ion, which then enters into a series of coordination complexes with excess thiourea. The latter, because of the presence of two amino groups, is more basic than a simple thioamide and is, thus, capable of forming mono salts with certain metallic cations. When using thiourea to complex cuprous ion, it is essential that an excess be provided to prevent the precipitation of the insoluble chloride mono salt. The sequence of possible reactions with copper ions is as follows:
Reduction of cupric ion by thiourea:

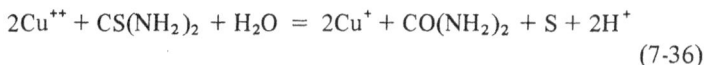

$$2Cu^{++} + CS(NH_2)_2 + H_2O = 2Cu^+ + CO(NH_2)_2 + S + 2H^+ \qquad (7\text{-}36)$$

Precipitation of insoluble thiourea mono salt (white):

$$Cu^+ + Cl^- + CS(NH_2)_2 = CuCS(NH_2)_2Cl \qquad (7\text{-}37)$$

Formation of soluble complex monomer:

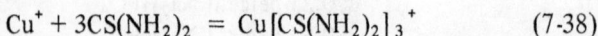

$$Cu^+ + 3CS(NH_2)_2 = Cu[CS(NH_2)_2]_3^+ \qquad (7\text{-}38)$$

Dimerization of the monomer:

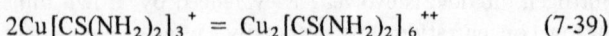

$$2Cu[CS(NH_2)_2]_3^+ = Cu_2[CS(NH_2)_2]_6^{++} \qquad (7\text{-}39)$$

Formation of a second monomer with a large excess of thiourea:

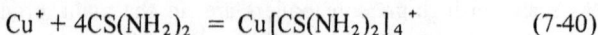

$$Cu^+ + 4CS(NH_2)_2 = Cu[CS(NH_2)_2]_4^+ \qquad (7\text{-}40)$$

Eq. (7-38) shows that a minimum of three moles of thiourea must be provided per mole of cuprous ion to keep the latter in solution, i.e., 3.6 lb of thiourea per pound of copper. Also, extra thiourea must be added to replace that oxidized to urea according to Eq. (7-36).

When magnetite in the presence of metallic copper is treated with a large excess of acid, copper is oxidized to cupric ion.

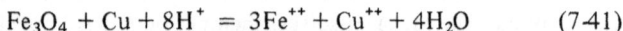

$$Fe_3O_4 + Cu + 8H^+ = 3Fe^{++} + Cu^{++} + 4H_2O \qquad (7\text{-}41)$$

If the acid solution contains sufficient thiourea, however, the otherwise unstable cuprous ion, formed as an intermediate product of the oxidation, is stabilized as the complex ions shown in the reactions in Eqs. (7-38), (7-39), and (7-40). In the absence of thiourea, any cuprous ion formed reverts to copper and cupric ion ($K = 1.6 \times 10^6$).[45]

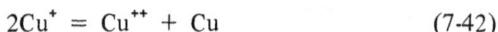

$$2Cu^+ = Cu^{++} + Cu \qquad (7\text{-}42)$$

In most instances of boiler fouling, the ratio of magnetite to copper is higher than the minimum demanded by Eq. (7-41).

The procedure for removing copper fouling is similar to that already prescribed using hydrochloric acid. The following formulation is typical:

> 5.0 percent hydrochloric acid
> 2.0 percent thiourea
> 0.4 percent proprietary corrosion inhibitor

After, filling, soaking, and rinsing, the cleaned metal is passivated as previously described. It is important that all copper be removed in the cleaning process, as a copper-plated steel surface cannot be passivated by the nitrite-phosphate solution recommended. Also, after cleaning, the boiler must be carefully examined for loosened foils or tubes of metallic copper left after an inefficient cleaning, for copper rapidly deteriorates steel as the result of potential differences between the dissimilar metals.

Wackenhuth and Richards[46] have investigated a three-stage treatment for removing copper, which they found to be so harsh that it produced deep pitting and gouging. In this method, the boiler to be cleaned is first filled with a 5 percent solution of inhibited hydrochloric acid at 150 F, then allowed to soak for 6 h. The acid solution is drained under nitrogen, then the boiler is filled with an ammoniacal solution containing 1 lb of sodium bromate per pound of copper to be removed. An excess of the oxidizing agent must be present to keep copper in the cupric state so that the reaction in Eq. (7-35) cannot occur.

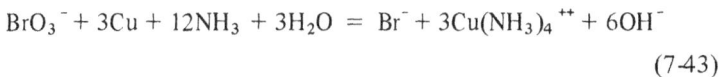

$$BrO_3^- + 3Cu + 12NH_3 + 3H_2O = Br^- + 3Cu(NH_3)_4^{++} + 6OH^-$$

(7-43)

The sodium bromate solution is heated to 150 F, allowed to soak for 6 h, then drained. Ammonium persulfate, $(NH_4)_2S_2O_8$, has been used as the oxidizing agent in this second step, but it too causes excessive pitting. Finally, a 5 percent solution of inhibited hydrochloric acid, containing 0.75 percent of thiourea, is heated to 150 F, and introduced into the boiler. After soaking for 6 h at 150 F, the solution is dumped under nitrogen, after which the boiler is passivated with the usual nitrite-phosphate solution (see pp. 67-68).

Other acids including 3 percent phosphoric at 200-212 F, 3 percent citric at 200-220 F, 10 percent sulfamic at 175 F, and 3 percent formic-hydroxyacetic acids at 200-220 F have been tried, but all of these require continuous circulation, and even then, they are much slower acting on boiler deposits than is dilute hydrochloric acid.[38] Other complexing agents that have been used are EDTA, diethylthiourea, and citrate; perborate is used as an oxidizing agent occasionally.

An excellent, though expensive, method for chemically cleaning boilers with an ammonium salt of EDTA is the subject of several patents.[47,48,49] A commercial solution containing about 40 percent by weight of triammonium EDTA and having a pH of 9.2 is used as the cleaning agent along with a corrosion inhibitor (either sodium 2-mercaptobenzothiazole or sodium gallate) at a concentration of 0.1-0.5 percent.[47] In practice, the boiler to be cleaned is cooled (or heated) until its pressure is about 100 psi. The steam drum is then partially drained to make room for the chelant solution, and the latter, together with the inhibitor, is injected into the boiler through any convenient port. The boiler is fired intermittently for 3-6 h so that its pressure varies from 50 to 150 psi (260-350 F). After cooling to 200 F, the boiler is vented, and compressed air is admitted through a lower blowdown or drain line; utility boilers having a capacity of 25,000-50,000 gal of water use about 2000 ft^3 of compressed air admitted over a period of 2-3 h. At the conclusion of the air-blowing, a single rinse with hot (180 F) condensate is sufficient.

Three phases comprise this method for removing boiler scale:

Oxide dissolution:

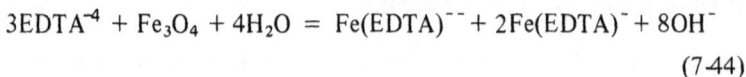

$$3EDTA^{-4} + Fe_3O_4 + 4H_2O = Fe(EDTA)^{--} + 2Fe(EDTA)^- + 8OH^-$$
$$(7-44)$$

Air oxidation of ferrous chelonate:

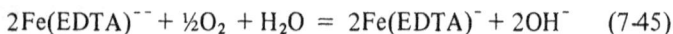

$$2Fe(EDTA)^{--} + \tfrac{1}{2}O_2 + H_2O = 2Fe(EDTA)^- + 2OH^- \qquad (7-45)$$

Oxidation of metallic copper by ferric chelonate:

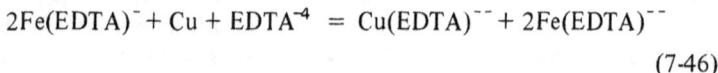

$$2Fe(EDTA)^- + Cu + EDTA^{-4} = Cu(EDTA)^{--} + 2Fe(EDTA)^{--}$$
$$(7-46)$$

Eq. (7-44) shows that 1 lb of iron requires 6.1 lb of $(NH_4)_3EDTA$, or 15.3 lb of a 40-percent solution of the salt. Similarly, Eq. (7-46) indicates that 1 lb of copper consumes 5.4 lb of the salt, or 13.5 lb of a 40-percent solution.

In the course of a cleaning operation, concentrations of dissolved iron, residual chelant, pH, and copper are checked frequently by methods given in Chapter 8. The amount of air needed to oxidize ferrous chelonate is found by measuring the oxidation potential of the solution with a saturated calomel electrode and a platinum electrode connected to a digital voltmeter. It is suggested in the original patent[48] that ferrous metal be used as the indicating electrode, but Latimer[45] has pointed out that iron electrodes polarize readily, and also, because of random surface strains, different pieces of steel yield large variations in potential measurements.

The foregoing cleaning procedure has a great many advantages over conventional methods of cleaning with acids.

1. The mildly alkaline solution is much safer to handle than acids.

2. Circulation of the cleaning solution is achieved by direct firing of the boiler rather than by an external pump.
3. Deposits are almost completely dissolved, which minimizes subsequent manual clean-up, as there is no precipitation during rinsing.
4. A single rinse with water is sufficient.
5. As the boiler metal is passivated during the air-blowing step, no additional application of a passivating solution is needed.
6. The exhausted solvent can be disposed of by incineration.
7. As a boiler can usually be cleaned in less than 24 h, marine boilers can be cleaned during a routine port call.
8. The corrosion rate of steel in the alkaline solvent is very low—0.3–0.5 mpd.

Before leaving the subject of chemical cleaning, brief mention of furnace and turbine cleaning is warranted. In most instances, furnaces are cleared of slag by mechanical chipping. Lee, et al.,[6] however, found that furnace slag is bound to the superheater and other tubes by water-soluble sulfates. Slag itself contains a great number of small interconnecting pores, which although impervious to water alone, are permeated by water containing 0.1–0.2 percent of a nonionic wetting agent. By saturating the slag in a furnace with water containing a wetting agent, and allowing it to stand for a day or two, part of the binding salt is dissolved, and the bulk of the slag becomes susceptible to removal by hydroblasting. This simple procedure, where applicable, saves considerable labor.

As mentioned in Section 7.3, cupric and cuprous oxides deposited in turbines severely impair the efficiency of the machines. Morris and Call[50] have reported their experience in flooding turbines with a patented copper solvent[51] containing 18 percent of ethylenediamine and 7.5 percent of hydroxylamine hydrosulfate. Both cuprous and cupric ions form stable coordination complexes with ethylenediamine.

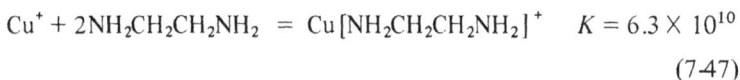

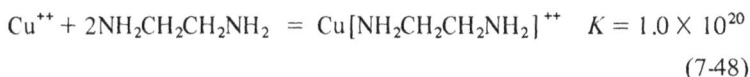

$$Cu^+ + 2NH_2CH_2CH_2NH_2 \ = \ Cu[NH_2CH_2CH_2NH_2]^+ \quad K = 6.3 \times 10^{10}$$

$$(7\text{-}47)$$

$$Cu^{++} + 2NH_2CH_2CH_2NH_2 \ = \ Cu[NH_2CH_2CH_2NH_2]^{++} \quad K = 1.0 \times 10^{20}$$

$$(7\text{-}48)$$

The role of hydroxylamine in the solvent is uncertain, but it readily decomposes to ammonium ion and nitrous oxide; as indicated by Eq. (7-49), the reaction is favored in alkaline solution.

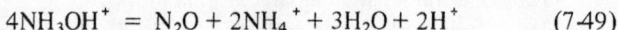

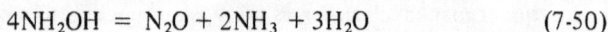

$$4NH_3OH^+ = N_2O + 2NH_4^+ + 3H_2O + 2H^+ \qquad (7\text{-}49)$$

$$4NH_2OH = N_2O + 2NH_3 + 3H_2O \qquad (7\text{-}50)$$

The majority of the methods of chemical cleaning discussed in this section are best left to experienced contractors who have proper equipment and who have made licensing arrangements for patented procedures. As with any purchased service, however, it is to the purchaser's advantage to have a good understanding of the work to be done and the methods to be used. For this reason this section has been included.

d. Reclamation of Ion-Exchange Resin

There is a tendency to discard ion exchange resins at the first sign of decreased exchange capacity, or in the case of anion exchange resins, of salt-splitting capacity. There are, however, several common contaminants that interfere with exchange reactions, but that can be removed by special procedures outlined below. The most common foulant of cation exchange resins is iron. It may be found mechanically entrapped in the exchanger as rust particles or as an insoluble floc, or it may be ionically attached to active exchange groups on the resin. Mechanically entrained particles of rust can be removed by vigorous backflushing, whereas gelatinous $Fe(OH)_3$ coats the resin beads and cannot be flushed out. Similarly, trivalent ferric ion that has been exchanged cannot be effectively displaced by the monovalent sodium ion. These tenacious forms of iron can be removed from cation exchange resin by filling the unit with 10 percent hydrochloric acid solution containing 0.1–0.2 percent of corrosion inhibitor and allowing it to soak for several hours. Sufficient acid is used to supply 1.5 gal of 30 percent hydrochloric acid per cubic foor of resin. After the soaking period, the acid is pumped into a vacuum truck for disposal. The unit is next rinsed, downflow, until most of the acid has been washed out, then backwashed for 30 min. Finally, the resin is regenerated twice with sodium chloride to ensure that all active exchange groups are in the sodium form.

Alternatively, iron can be removed by treating the resin with sodium hydrosulfite, $Na_2S_2O_4$, at a dosage of 1 lb per cubic foot of resin. This salt reduces ferric to ferrous ion, which is much less firmly held by the resin and is, thus, more easily displaced by sodium ion. The cleaning solution is prepared by dissolving in hot water 1 lb of $Na_2S_2O_4 \cdot 2H_2O$ and 5 lb of

NaCl per cubic foot of resin to be treated. The solution is then pumped into the ion exchange unit, while agitating the resin and solution with nitrogen. After soaking for several hours, the reducing solution is drained and the unit is rinsed until no odor of sulfur dioxide is detectable, after which the resin is regenerated with sodium chloride.

As sodium hydrosulfite decomposes readily to thiosulfate, sulfurous acid, and sulfur,[52] the solution should be used immediately after being prepared.

$$2S_2O_4^{--} + H_2O = 2HSO_3^- + S_2O_3^{--} \qquad (7\text{-}51)$$

$$2H^+ + S_2O_3^{--} = SO_2 + S + H_2O \qquad (7\text{-}52)$$

To avoid oxidation, sodium hydrosulfite solution must not be agitated with air.

$$S_2O_4^{--} + O_2 + H_2O = HSO_4^- + HSO_3^- \qquad (7\text{-}53)$$

Also, the solid hydrated salt must be protected from exposure to air, as it is readily oxidized to sodium pyrosulfite.

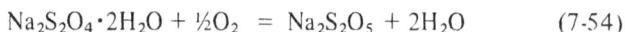

$$Na_2S_2O_4 \cdot 2H_2O + \tfrac{1}{2}O_2 = Na_2S_2O_5 + 2H_2O \qquad (7\text{-}54)$$

Iron, introduced in sodium hydroxide, can mechanically foul anion exchangers and also may be bound as anionic complexes, e.g., ferric citrate, $Fe(C_6H_5O_7)_2^{---}$, that are not displaced by hydroxyl ion. Before attempting to remove iron, the resin should be regenerated in the usual way to eliminate silicate. This step is followed by the successive application of 10 percent sodium chloride and a dilute solution of hydrochloric acid (5 percent). The unit is then rinsed, backwashed, and regenerated with sodium hydroxide. This treatment also dissolves calcium carbonate and magnesium hydroxide, both of which are precipitated in anion exchange resin if hard water is used to make up sodium hydroxide solution, or for rinsing.

Oils, which are not removed by either hydrogen or sodium ions in normal regeneration, often foul cation exchange resins. To remove oily material, it is usually sufficient to pass a warm 0.1-percent solution of a nonionic wetting agent through the bed of contaminated resin. Should heavy oils be encountered, it may be necessary to wash the resin with dilute sodium hydroxide containing a wetting agent, in which case the resin should first be regenerated by hydrogen ion to displace any cations that are precipitated by hydroxyl ion. Other organic matter such as tannins, phenols, and humic acids that are present in surface waters[53,54] are also removed fairly completely by the alkaline detergent solution, but oxidative degradation products of cation exchange resins themselves are irreversibly adsorbed on anion resins, permanently decreasing their exchange capacity.

Because of the tendency of organic matter to concentrate on the surface of ion exchange beds, bacteria often grow there in large numbers. *Pseudomonas* is especially common, often at concentrations of 6–10 million per gram of resin. If exchangers are regenerated every 3–4 days, most bacteria are removed in the process. When they begin to proliferate, however, they can be controlled by the application of formaldehyde, hypochlorite, or chloromelamine (chlorocyanuramide).

Further details of all of the procedures outlined here are available from the manufacturers of ion-exchange resins.

REFERENCES

(1) McCoy, J. W. Apr., 1957. *The systematic analysis of deposits from oil-fired furnaces.* ASTM Bulletin, No. 221, 59.

(2) McCoy, J. W. 1969. *Chemical analysis of industrial water.* New York: Chemical Publishing.

(3) Beach, L. K. and Shewmaker, J. E. 1957. The nature of vanadium in petroleum. *Ind. Eng. Chem.* 49:1157.

(4) Jones, M. C. K. and Hardy, R. L. 1952. Petroleum ash components and their effect on refractories. *Ind. Eng. Chem.* 44: 2615.

(5) Salooja, K. C. Jan., 1973. Burner fuel additives. *Combustion* 44(7):21.

(6) Lee, G. K., Mitchell, E. R., Grimsey, R. G., and Hopkins, S. E. 1964. An investigation of fuel-oil additives to prevent superheater slagging in naval boilers. *Proc. Amer. Power Conf.* 26: 531.

(7) Anonymous, 1969. Improving combustion efficiency. *J. Fuel and Heat Technol.* 16(6):24.

(8) Kukin, I. 1973. Additives can clean up oil-fired furnaces. *Environ. Sci. Technol.* 7:606.

(9) Kukin, I. Sept. 24, 1974. Combustion control by additives introduced in both hot and cold zones. U.S. Patent No. 3,837,820.

(10) Ramsdell, R. G., Noon, W. J., and Newman, C. L. Mar. 26–28, 1973. *Application of fuel oil additives for utility boilers.* presented before the Canadian Electrical Association, Toronto, Ontario.

(11) Glasstone, S. and Lewis, D. 1960. *Elements of physical chemistry.* Princeton: Van Nostrand.

(12) Holmes, J. A. and Jacklin, C. Jan., 1944. Experimental studies of boiler scale at 800 psi. *Combustion* 15(7):35.

(13) Miyakawa, M., Mizutani, T., Yagi, R., Sakae, Y., and Sirakawa, S. Sept., 1972. Estimation of deposit formation for boiler tube inner surface. *Combustion* 44(3):6.

(14) McCoy, J. W. 1974. *The chemical treatment of cooling water.* New York: Chemical Publishing.

(15) McCoy, J. W. 1980. *The microbiology of cooling water.* New York: Chemical Publishing.

(16) Rice, J. K. 1960. Deposit identification—first step toward understanding a water problem, in *Amer. Soc. Testing Materials.* Special Technical Publication No. 256. Symposion on identification of water-formed deposits.

(17) Colacito, P. M. Sept., 1971. Polarizing microscope identifies water-formed deposits. *Research/Development* 22(9):22.

(18) McCoy, J. W. 1954. Systematic analysis of boiler deposits. *J. Amer. Water Works Assoc.* 46:903.

(19) Howell, F. W. and McConomy, T. A. 1966. Maintaining turbine capability through boiler water purification. *Proc. Amer. Power Conf.* 28:808.

(20) Straub, F. C. and Grabowski, H. A. 1945. Silica deposition in steam turbines. *Trans. ASME* 67:309.

(21) Pocock, F. J. and Stewart, J. F. 1963. The solubility of copper and its oxides in supercritical steam. *J. Eng. Power* 85-A: 33.

(22) Noll, D. E. 1964. Factors that determine treatment for high-pressure boilers. *Proc. Amer. Power Conf.* 26:753.

(23) Neat, F. U. Mar., 1961. The relation between boiler cleanliness and feedwater. *Combustion* 32:35.

(24) Garrels, R. M. and Christ, C. L. 1965. *Solutions, minerals, and equilibria.* New York: Harper and Row.

(25) Long, N. A. 1966. Recent operating experiences with stainless steel condenser tubes. *Proc. Amer. Power Conf.* 28:798.

(26) Klein, H. A., Lux, J. A., Riedel, W. L., Noll, D. E., and Phillips, H. 1971. A field survey of internal boiler tube corrosion in high pressure utility boilers. *Proc. Amer. Power Conf.* 33: 357.

(27) Uhlig, H. H. 1963. *Corrosion and corrosion control.* New York: John Wiley & Sons.

(28) Clarke, F. E. and Ristaino, A. J. 1962. *New clues in the boiler tube pitting puzzle.* First Intern. Congr. Metallic Corrosion. 1st London, Engl. 1961. 403–7.

(29) Noll, D. E. 1958. Limitations on chemical means of controlling corrosion in boilers. *Corrosion* 14:541t.

(30) Ames, W. C. and Lux, J. A. 1967. An experimental investigation of hydrogen damage in boiler tubing. *Proc. Amer. Power Conf.* 29:763.

(31) Van Brunt, J. July, 1945. Tube failures in water-tube boilers. *Combustion* 17:32.

(32) Goodstine, S. L. and Kurper, J. L. May, 1973. Corrosion and corrosion product control in the utility boiler-turbine cycle. *Combustion* 44:6.

(33) Dean, J. A. 1973. *Lange's handbook of chemistry.* 11th ed., New York: McGraw-Hill.

(34) Herman, K. W. and Gelosa, L. R. Apr., 1973. Water treat-

ment for heating and process steam boilers. *Power Eng.* 77(4): 54.

(35) Edwards, J. C. and Rozas, E. A. 1961. Boiler scale prevention with EDTA chelating agents. *Proc. Amer. Power Conf.* 23:575.

(36) Edwards, J. C. and Merriman, W. R. 1963. Use of chelating agents for continuous internal treatment of high pressure boilers. Proc. Twenty-Fourth Annual Water Conf., *Eng. Soc. W. Pa.* 24:35.

(37) Wackenhuth, E. C. and Richards, H. E. 1966. Recent experiences in chemically cleaning a modern boiler. *Proc. Amer. Power Conf.* 28:785.

(38) Atwood, K. L. Sept., 1970. Solvent selection for preoperational and operational cleaning of utility boilers. *Combustion* 42(3):16.

(39) Anderson, J. D., Hayman, E. S., Jr., and Rodzewich, E. A. Nov. 16, 1976. Acid inhibitor composition and process in hydrofluoric acid chemical cleaning. U.S. Patent No. 3,992,313.

(40) Mannich, C. and Krösche, W. 1912. Ueber ein kondensationsprodukt aus formaldehyd, ammoniak und antipyrin. *Arch. Pharm.* 250:647.

(41) Rice, J. K. July, 1961. Repeated acid cleaning of boilers. *Combustion* 33(1):45.

(42) Greenberg, S. 1966. Factors that must be considered for successful chemical cleaning as experienced in naval boilers. *Proc. Amer. Power Conf.* 28:818.

(43) Martin, R. C. and Abel, W. T. Nov. 8, 1960. Copper and iron containing scale removal from ferrous metal. U.S. Patent No. 2,959,555.

(44) Engle, J. P. Aug., 1961. The merit of a single stage solvent for simultaneous removal of copper and iron oxides. *Combustion* 33(2):41.

(45) Latimer, W. M. 1952. *The oxidation states of the elements and their potentials in aqueous solutions.* 2nd ed. Englewood Cliffs, New Jersey: Prentice-Hall.

(46) Wackenhuth, E. C. and Richards, H. E. 1966. Recent experiences in chemically cleaning a modern boiler. *Proc. Amer. Power Conf.* 28:785.

(47) Lesinski, C. A. Mar. 7, 1967. Scale removal, ferrous metal passivation and compositions therefor. U.S. Patent No. 3,308,065.

(48) Teumac, F. N. Nov. 26, 1968. Passivation of ferrous metal surface. U.S. Patent No. 3,413,160.

(49) Harriman, L. W., Muehlberg, P. E., and Teumac, F. N. Apr. 15,

1969. Removal of copper containing incrustations from ferrous surfaces. U.S. Patent No. 3,438,811.
(50) Morris, E. B. and Call, R. G. *Chemical cleaning of turbines for
removal of copper oxide deposits.* ASME Paper 64-WA/BFS-1.
(51) Call, R. G. Oct. 10, 1961. Cleaning composition and a method
of its use. U.S. Patent No. 3,003,970.
(52) Yost, D. M. and Russell, H., Jr. 1944. *Systematic inorganic
chemistry.* New York: Prentice-Hall.
(53) Wilson, A. L. 1959. Organic foulant of strongly basic anion
exchange resins. *J. Appl. Chem.* 9:352.
(54) Frisch, N. and Kunin, R. 1960. Organic fouling of anion exchange resins. *J. Amer. Water Works Assoc.* 52:875.

Chapter 8.

Analytical Methods

In treating boiler water it may be said that the accurate control of operating variables is of much greater importance than the specific method of treatment. Thus, regardless of the program of treatment selected, poor results will be realized if the program is not properly carried out. Included in the first section of this chapter are a dozen or so analytical methods designed for use in boiler plants. Without pretending to the precision of standard procedures,[1,2] they are considerably more accurate than the average test kit that depends on drop counting, color comparators, or other inexact measurement. It is well worth the expense to equip the boiler plant with burets, a photoelectric colorimeter, a pH meter, and other standard laboratory equipment, rather than prepackaged kits with unidentified proprietary chemical reagents. The boiler operator should also be provided with a sampling and testing schedule, similar to Table 6.21, and a list of recommended operating concentrations, as in Table 6.12.

In addition to the control methods intended for boiler plant operators in the first section, several special nonroutine laboratory procedures are included in this chapter. Among them are methods for analyzing chemical cleaning solutions, evaluating ion exchange resins, and assaying neutralizing amines. In general, these procedures should be carried out in an analytical laboratory.

8.1 BOILER PLANT CONTROL METHODS

For obvious reasons, it is highly desirable to make the facilities for chemical testing as convenient as possible for the operator. Permanent setups of self-filling burets should be provided, one for each standard titrating solution used. A suitable instrument for measuring color is essential—either a photoelectric colorimeter* or a relatively inexpensive spectrophoto-meter**—as well as a reliable pH meter and a conductivity bridge with appropriate dip cells. Instructions are provided in the procedures for preparing special reagents, which, in general, are either purchased from a laboratory supply house or else provided by a company laboratory. Care must, of course, be taken to renew unstable reagents regularly and to avoid cross-contamination of reagents.

In order to obtain a representative portion of boiler water for analysis, the continuous blowdown line should be fitted with a sampling valve and connection ahead of the flash tank. The sampling coil should be of stainless steel—as short as possible—and equipped with a cooler to prevent flashing. Samples from intermittent blowdown connections on the mud drum or water wall headers are unsatisfactory, as the concentration of dissolved solids in these sections is lower than in the steam drum. Also, sludge tends to accumulate in the bottom parts of a boiler. When drawing a sample, the blowdown valve itself should be flushed as well as the sampling valve to purge insoluble material from the collection header and piping and to free both valves from deposits. The following sequence of operations is suggested.

1. Open the sampling valve wide and note the micrometer setting on the continuous blowdown valve.
2. Open the blowdown valve wide for about a minute, then close it completely.
3. Carefully open the blowdown valve to the original setting.
4. Regulate the flow of water through the sampling valve to ensure adequate cooling.
5. Rinse a clean sampling container a few times, then collect a sample of boiler water.

In interpreting analytical results both the operator and the plant manager should bear in mind that at high steaming rates sodium salts of phosphate, silicate, and sulfate are prone to precipitate on steam-generating surfaces, then return to solution when the steam load is reduced. This solubility effect, caused by a concentrating film, is commonly called

---

* For example, the Hellige Aqua Analyzer, available from Hellige, Inc., 877 Stewart Avenue, Garden City, New York.
** the Spectronic 20, available from Bausch and Lomb, 635 St. Paul Street, Rochester, New York 14602.

"hideout;" this phenomenon is discussed in more detail on page 173. The net effect is that lower concentrations of sulfate, phosphate, and silicate are likely to be indicated at high steam loadings than at lower rates.

a. Determination of Alkalinity

The determination of various types of alkalinity by titration with a standard acid solution is a simple and important method for controlling the operation of a number of water systems. By measuring the alkalinities indicated by phenolphthalein and either methyl orange or a mixed indicator (P and M values), it is possible to calculate the approximate concentrations of hydroxide, carbonate, and bicarbonate in most water samples (see Table 5.4). These values are significant in controlling corrosion and preventing scaling in boilers, and in the operation and control of various processes for treating make up water for boilers. The use of equivalent ppm $CaCO_3$ as a unit of alkalinity and hardness facilitates the calculation of proper dosages in a number of operations in water treating, and simplifies the comparison of analyses of different waters.

When an alkaline sample is titrated to phenolphthalein to determine the P value, all free hydroxide, half of the carbonate, and about one-third of any phosphate present are titrated. The hydrolysis of the latter two ions produces hydroxyl ion, making these salts alkaline to the phenolphthalein indicator. The reactions are as follows:

$$CO_3^{--} + H_2O = HCO_3^- + OH^- \tag{8-1}$$

$$PO_4^{---} + H_2O = HPO_4^{--} + OH^- \tag{8-2}$$

The hydroxyl ion formed by these reactions is titrated with standard acid to the phenolphthalein end point. The pH at the first perceptible pink color of phenolphthalein is 8.3. Thus, CO_3^{--} is titrated to HCO_3^-, but PO_4^{---} is titrated past HPO_4^{--} and, because the change of color occurs in a buffered region, the visual end point is poor in the presence of much phosphate.

To obtain the M value, a mixed indicator is added and the titration is continued to the end point occurring at about 4.6 pH. Bicarbonate ion is thus titrated to H_2CO_3 and HPO_4^{--} is titrated to $H_2PO_4^-$.

To determine the concentration of free hydroxyl ion, as distinguished from that formed by hydrolysis, neutral barium chloride solution is added to precipitate carbonate and phosphate.

$$Ba^{++} + CO_3^{--} = BaCO_3 \qquad\qquad (8\text{-}3)$$

$$3Ba^{++} + 2PO_4^{---} = Ba_3(PO_4)_2 \qquad\qquad (8\text{-}4)$$

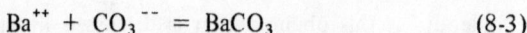

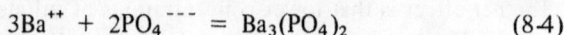

The solution is then titrated to the end point of phenolphthalein with standard acid to give a B value representing the free hydroxyl ion present. Since the pH of a saturated solution of barium carbonate is 8.6, the pink color of the indicator recurs. Also, the precipitation of phosphate is not quite complete: a solution containing 250 ppm of PO_4^{---} has a B value of about 0.5. The same solution shows a P value of 8 and an M value of 21. Concentrations of phosphate encountered in water treating are much lower than this, however, so phosphate seldom causes appreciable interference.

Phenolphthalein Alkalinity – P Value

Procedure: Transfer a 100-ml sample to a Coors No. 3A porcelain casserole and add 4 drops of phenolphthalein indicator.

Titrate to the phenolphthalein end point with standard 0.02 N sulfuric acid and record the titration.

Reserve the solution for the determination of M value, if required.

$$P \text{ value, ppm } CaCO_3 = \text{ml titration} \times 10$$

Methyl Orange Alkalinity – M Value

Reagent: Mixed Indicator. Dissolve 0.02 g of Methyl Red (Eastman No. 431) and 0.1 g of Bromcresol Green (Eastman No. 1782) in 100 ml of Formula 30 alcohol.

Procedure: To the solution used for determining the P value add 4 drops of mixed indicator and continue the titration to a definite reddish tinge.

Record the total titration and discard the solution.

$$M \text{ value, ppm } CaCO_3 = \text{ml total titration} \times 10$$

Hydroxide Alkalinity – B Value

Procedure: Transfer a 60-ml sample to a Coors No. 3A porcelain casserole or other suitable container.

Add 4 drops of phenolphthalein indicator and 10 ml of neutralized 10 percent $BaCl_2$ solution.

Titrate to the phenolphthalein end point with standard 0.02 N sulfuric acid, disregarding any recurrence of the pink color on standing.

$$B \text{ value, parts } CO_3^{--}/100,000 = \text{ml titration}$$
$$B \text{ value} \times 16.6 = \text{ppm } CaCO_3$$
$$B \text{ value} \times 13.3 = \text{ppm } NaOH$$

b. Determination of Chloride

The well-known Mohr method[3] is prescribed here for determining chloride in boilers and boiler feed waters. This procedure is generally used to control blowdown, but results may be misleading because of the rather poor accuracy at low concentrations. It should be realized that the concentration of chloride in boiler feed water varies considerably in installations using cationic softeners regenerated with sodium chloride. Immediately after regenerating a softener with brine and returning it to service, the concentration of chloride in the feed water rises for a short while. If the feed water is sampled during this period, the blowdown ratio will be erroneous. If possible, a 24-h composite sample of boiler feed water should be used, prepared by adding 80-100 ml of feed water to a quart bottle every 2 h.

The Mohr method for chloride is based on the differential precipitation of chloride ion and chromate ion, the latter serving as the indicator of the end point by precipitating as the brick-red Ag_2CrO_4 after all of the chloride has precipitated as the less-soluble AgCl. As the sensitivity of the indicator is markedly decreased in acidic solutions, the pH should be > 7, but < 10.5. For convenience, a normality of silver nitrate is used so that for a 100-ml sample size it is only necessary to multiply the buret reading by 10 to obtain ppm of NaCl directly. As this solution is dilute, a fairly large indicator blank is obtained, making the determination of small concentrations of sodium chloride rather inaccurate.

A study by Sheen and Kahler[4] has shown that the Mohr titration is unaffected by normal amounts of sulfate, alkalinity, hardness, phosphate, silicate, and iron. Sulfite interferes to some extent, but it is eliminated by pretreating the sample with a small amount of dilute hydrogen peroxide. When neutralizing alkaline samples, care must be taken to discharge all of the color of phenolphthalein. After boilers have been shut down for cleaning or inspection, blowdown water samples often contain finely divided ferric oxide for several days after the boiler has been returned to service. This material interferes in the Mohr titration by imparting an orange cast to the solution; such samples should be filtered before attempting to titrate chloride.

Sodium Chloride in Water

Reagents: Standard Silver Nitrate Solution (0.0171 N; 1 ml = 1.00 mg NaCl).   Catalog No. HL 2830-4, Hartman-Leddon Company, 60th and Woodland Avenue, Philadelphia, Pa. 19143.
Potassium Chromate Indicator. Dissolve 5 g of K_2CrO_4 in 100 ml of water.
Hydrogen Peroxide Solution. Dilute 10 ml of 30 percent H_2O_2 to 100 ml.
Procedure: Transfer 100 ml of a filtered sample to a Coors No. 3A porcelain casserole. If sulfite is present add 10 ml of 3 percent H_2O_2.

Add 1-2 drops of phenolphthalein indicator and, if the solution is pink, add 0.02 N H_2SO_4 slowly until the solution is colorless.

If the solution is initially colorless, add about 0.2 g of solid $NaHCO_3$ to adjust the pH to about 8.

Add 1.0 ml of 5 percent K_2CrO_4 indicator solution and titrate with 0.0171 N $AgNO_3$ solution to a faint orange end point.

Make a blank determination, substituting distilled water for the sample. This will amount to about 0.6 ml.

$$\text{ppm NaCl} = (\text{titr.} - \text{blank titr.}) \times 10$$

$$\text{ppm Cl}^- = \text{ppm NaCl} \times 0.6$$

c. Determination of Dissolved Solids

Dissolved solids are customarily determined by weighing the dried residue obtained after evaporating a sample of water. If the residue is dried at 105 C, constant weight is attained only slowly and some of the salts retain water of crystallization. Drying the residue at 180 C gives results in approximate agreement with those obtained by adding together all of the components determined individually. In the operation of boilers, the control of dissolved solids is important in preventing mechanical carryover and for ensuring the purity of steam. The determination of dissolved solids by measuring the specific conductance of a sample of water is based on the fact that the more concentrated the ions in a solution the greater is its conductivity. The specific conductance of a solution is roughly proportional to the residue obtained by drying at 180 C, and, if desired, a ratio can be established that can be used to convert specific conductance to dissolved solids.

Referring to Table 6.22, it is noted that the specific conductance of NaOH is much greater than those of equal concentration of other salts; this is because of the high ionic mobility (20.5×10^{-14} cm/s) of the hydroxyl ion. Maguire and Polsky[5] have recommended the addition of a measured excess of gallic acid, 3,4,5-trihydroxybenzoic acid, to samples from boilers before determining specific conductance. This neutralizes the free hydroxyl ion and forms an equivalent amount of gallate ion; the conductance of sodium gallate solutions approximates that of other boiler salines.

As the acid dissociation constant of gallic acid is 4×10^{-5}, the acid itself increases the conductivity of distilled water and therefore should be used only in alkaline samples containing rather high concentrations of dissolved solids. The amount of gallic acid specified in the following procedure produces a specific conductance of about 125 μmhos when dissolved in 100 ml of distilled water. In Table 8.1 are listed the specific conductances of two sets of boiler water, the specific conductances after adding gallic acid, and the dissolved solids determined by evaporation and drying at 180 C.

TABLE 8.1

Conductance and Dissolved Solids

	Boiler	μmhos	+ Gallic acid (μmhos)	Dissolved solids (ppm)	Factor*
900 psi	1	1350	980	834	0.85
	2	1380	1000	870	0.87
	3	1550	1170	1030	0.88
	4	1650	1220	1062	0.87
	5	1500	1150	1027	0.89
					Avg. 0.87
450 psi	1A	1600	960	871	0.91
	2A	1600	900	852	0.95
	3A	1650	925	870	0.94
					Avg. 0.93

* Gravimetric solids divided by μmhos (with gallic acid).

Note that the ratio of gravimetric solids to specific conductance is reasonably constant at each pressure. The ratio must be determined for each type of water in which it is desired to relate total dissolved solids to specific conductance. Specific conductance is given by the expression

$$\bar{L} = dL/A \qquad (8\text{-}5)$$

where d is the distance between electrodes (cm), A is the area of the electrodes (cm^2), and L is the conductivity. Conductivity bridges that read specific conductance directly when fitted with a suitable dip cell are commercially available. The term d/A is called the cell constant; by selecting cells with suitable cell constants, specific conductance can be measured over a wide range. For instance, a constant of 0.1 is suitable for measurements in the 1-500 μmho range, while a constant of 2.0 is used for the 20-20,000 μmho range, i.e., twenty times greater.

Conductometric Determination of Dissolved Solids

Instrument: Conductivity Bridge. Any suitable commercial model equipped with appropriate dip cells, temperature compensator, range selector, and null point indicator.

Procedure: Prepare the conductivity bridge and dip cells for the measurement. If a dual-range instrument is used, be sure that the correct dip cell is used for the range selected. The cell with the lower cell constant is used for the lower range; the leads and terminals are usually color coded.

If the specific conductivity is desired, take the temperature of the water sample, immerse the dip cell and move it up and down to dislodge air bubbles. Set the temperature compensator to the thermometer reading, and adjust the bridge to the null point.

Report the specific conductance as read from the dial of the instrument.

If the total dissolved solids concentration is desired, add 0.2 g of gallic acid, stir to dissolve, immerse the dip cell, and measure the specific conductance as above. Multiply the conductance obtained by a previously determined factor, and report the concentration of total dissolved solids. (As mentioned in the discussion, this method is applicable only to boiler waters with high alkalinity.)

d. Determination of EDTA Chelant

When using EDTA, or other chelant, to prevent scaling in boilers by

alkaline earth elements, it is important that the addition of chelant be proportioned to the concentration of calcium and magnesium in the feed
water. The accurate determination of hardness values less than 1 ppm
$CaCO_3$, however, is difficult in a laboratory and impracticable in a boiler
plant. Furthermore, hardness varies in feed water because of leakage from
softeners, contamination of condensate by raw water, and other sources
of hardness. For this reason, it is convenient to control EDTA dosage by
measuring the concentration of residual (uncombined) chelant in the boiler
water. This is easily done by titrating a sample of blowdown water with a
standard solution of magnesium chloride, which combines readily with
residual chelant. This procedure is essentially the reverse of the titration
of total hardness described in the next section. As the color change of
Eriochrome Black T indicator from red to blue is much sharper than from
blue to red, it is convenient to add an excess of standard magnesium solution, then back-titrate with standard EDTA solution.

Total chelant is determined colorimetrically by adding a measured
excess of zirconium ion, which because of its large formation constant
($K_f = 2.5 \times 10^{19}$), displaces calcium and magnesium from their EDTA
chelonates. The sample size is so chosen that there is an excess of zirconium ion, which forms a red complex with xylenol orange. By measuring
both residual and total chelant it is possible to calculate the average total
hardness in the feed water. Suppose, for example, that

$$Total\ EDTA\ =\ 15\ ppm$$

$$\underline{Residual\ EDTA\ =\ \ 3\ ppm}$$
$$Combined\ EDTA\ =\ 12\ ppm$$

As 380 mg of Na_4EDTA chelates 100 mg of $CaCO_3$, 1 ppm Na_4EDTA
chelates 0.26 ppm $CaCO_3$. Therefore, 12 ppm EDTA is combined with
3.16 ppm of $CaCO_3$. If the boiler is being operated at, say 5 percent blowdown, then the average total hardness in the feed water is $(3.16 \times 0.05) =$
0.16 ppm $CaCO_3$.

Residual EDTA in Boiler Water

Reagents: Buffer Solution. Dissolve 40 g of $Na_2B_4O_7 \cdot 10H_2O$ in 800 ml
of water. Dissolve 10 g of NaOH, 5 g of $Na_2S \cdot 9H_2O$, and 10 g of KNa-
$C_4H_4O_6 \cdot 4H_2O$ in 100 ml of water. Combine the two prepared solutions,

dilute to 1000 ml, and store in a stoppered bottle. This solution is stable for about one month.

Indicator. Mix 0.2 g of Eriochrome Black T (Eastman No. P6361) with 80 g of NaCl, grind to 40–50 mesh, and store in a dark bottle.

Standard Magnesium Solution. Carefully dissolve 64 mg of magnesium ribbon in 50 ml of water containing 1 ml of HCl. Evaporate to dryness, then dilute to 1000 ml in a volumetric flask. (1 ml is equivalent to 1.00 mg of Na_4EDTA.)

Standard EDTA Solution. Dissolve 0.979 g of disodium ethylenediamine-tetraacetate dihydrate in water and dilute to 1000 ml in a volumetric flask. (1 ml = 1.00 mg Na_4EDTA.)

Procedure: Transfer 100 ml of boiler water to a Coors No. 3A porcelain casserole. Add 10.0 ml of standard magnesium solution, followed by 1.0 ml of buffer solution and 0.2 g of Eriochrome Black T indicator, stirring between additions.

If the solution is not red at this point, add an additional 5.0-ml portion of standard magnesium solution.

Titrate slowly with standard EDTA solution until the color changes from wine-red to clear blue, and record the buret reading.

$$\text{Residual chelant, ppm } Na_4EDTA = (\text{ml Mg sol.} - \text{ml EDTA sol.}) \times 10$$

Total EDTA in Boiler Water

Reagents: Standard EDTA Solution. Prepare as in the above procedure. (1 ml = 1.00 mg Na_4EDTA).

Xylenol Orange Solution. Dissolve 0.20 g of xylenol orange (Eastman No. 9964) in 85 ml of HCl. Dissolve 25 g of $NH_2OH \cdot HCl$ in 100 ml of water. Combine the prepared solutions and dilute to 250 ml in a volumetric flask. Allow to stand overnight, then filter, if necessary.

Zirconyl Chloride Stock Solution. Dissolve 1.800 g of $ZrOCl_2 \cdot 8H_2O$ (Fisher Catalog No. Z-80) in 500 ml of water containing 65 ml of HCl, and dilute to 1000 ml. (1 ml = 0.50 mg Zr.)

Dilute Zirconyl Chloride Solution. Transfer 10.0 ml of the stock solution and 5 ml of HCl to a 250-ml volumetric flask; dilute to the mark with water. Prepare a fresh solution, as required. (1 ml = 0.020 mg Zr.)

Standardization: Dilute 10.0 ml of standard EDTA solution to 100 ml to produce a solution containing 0.10 mg Na_4EDTA/ml. Transfer 1.0, 2.0, 3.0, and 4.0 ml of dilute EDTA standard corresponding to 0.10, 0.20, 0.30, and 0.40 mg of Na_4EDTA to a series of 50-ml volumetric flasks and include a blank.

Bring the total volume of each flask to about 35 ml, then add, while swirling, 5.0 ml of xylenol orange. Dilute the blank only to 50 ml.

Add 5.0 ml of dilute zirconyl chloride solution to each of the flasks containing aliquots of the dilute standard EDTA solution. Dilute to volume and mix thoroughly.

Allow the solutions to stand for 30 min, then set the transmittance of the blank at 100 percent at a wavelength of 535 nm. Measure the transmittances of the standards and plot the results against the corresponding mg of Na_4EDTA on semilogarithmic graph paper.

Procedure: Filter the sample if it is not clear, transfer a 25-ml portion to a 50-ml volumetric flask, and dilute to about 35 ml.

Prepare a blank by transferring 5.0 ml of xylenol orange solution to a 50-ml volumetric flask, and diluting to the mark with water.

To the sample, add 5.0 ml of xylenol orange solution, 5.0 ml of dilute zirconyl chloride solution, and dilute to the mark with water.

Allow the flasks to stand for 30 min, set the transmittance of the blank solution at 100 percent, measure the transmittance of the sample at 535 nm, and read the corresponding mg of Na_4EDTA from the prepared calibration curve.

$$ppm\ Na_4EDTA\ =\ mg\ Na_4EDTA \times 10^3/ml\ sample$$

e. Determination of Hardness

Over the years many different methods have been proposed for determining the total concentration of calcium and magnesium in a variety of waters including raw water, condensates, process waters, feed water for boilers, and effluents from treating equipment such as softeners and units for removing silica. The most convenient method for determining hardness is the colorimetric titration of Schwarzenbach, et al.,[6] based on the sequestering reaction of disodium dihydrogen ethylenediaminetetraacetate with calcium and magnesium ions using Eriochrome Black T to detect the end point.[7] The titration is carried out in a solution buffered at pH 10. Schwarzenbach originally used ammonium chloride and ammonia, the buffer still preferred by some authors. A study by Diehl, et al.,[8] however, shows that water containing polyphosphate cannot be satisfactorily titrated for total hardness using the ammonia-ammonium chloride system. Betz and Noll[9] found that a tetraborate-hydroxide buffer containing sodium sulfide, as recommended by Marcy,[10] gave correct results in the presence of as much as 25 ppm of polyphosphate; 50 ppm caused slightly low results. Sodium sulfide is included to prevent the interference of copper and manganese.

This buffer has been criticized by Diehl, et al.[8] as giving broad indistinct end points. In addition, sulfide does not keep well, the amount of buffer added is critical, and, in the presence of iron, the black precipitate of ferrous sulfide disturbs the end point. Despite these objections, the borate buffer has been adopted for a standard method,[11] with the addition of a small amount of rochelle salt, potassium sodium tartrate, to prevent interference by iron. It gives satisfactory results with the majority of water samples, and the lack of interference by moderate amounts of polyphosphates is a definite advantage.

Slightly ionized calcium and magnesium compounds are formed with Eriochrome Black T, the most stable of which is that with magnesium. As magnesium EDTA is more dissociated than the calcium chelonate, calcium is sequestered first by EDTA. As the titration proceeds, magnesium is withdrawn from the dye, and at the end point the color changes sharply to blue. At pH 10 the magnesium dye is wine-red and the dye itself is blue.

A difficulty that sometimes arises in this method is caused by high concentrations of bicarbonates. When the buffer solution is added to such samples, $CaCO_3$ precipitates. If the water is cold $CaCO_3$ dissolves rapidly during the titration and correct results are obtained. If the water is warm (above 65 C), however, the precipitate dissolves but slowly and low results or a constantly recurring end point may be obtained. This generally happens if the concentration of bicarbonate exceeds 250 ppm $CaCO_3$. This difficulty can be avoided by adding a small amount (1 ml) of 2N HCl before the buffer solution is added to the sample.

A titration of calcium with EDTA using ammonium purpurate was proposed by Schwarzenbach[6] and tentatively described by Connors,[12] who also recommended the use of sodium diethyldithiocarbamate to prevent the interference of copper ions often introduced into water supplies in the treatment of reservoirs with copper sulfate. Betz and Noll[9] investigated the method and with further development found it satisfactory. No buffer is used, but the pH is raised by the addition of 1N NaOH. Ammonium purpurate (also called murexide and acid ammonium purpurate) is used as the indicator, with a color change from salmon-pink to orchid. One unfamiliar with the final color would find it helpful to compare the color with that obtained in distilled water. Since the indicator is unstable in aqueous solutions, it is always used as the dry powder dispersed in solid sodium chloride.

Total Hardness – TH

Reagents: Buffer Solution. Prepare as in Section 8.1d above.
Indicator. Prepare as in Section 8.1d above.
Standard Calcium Solution. Carefully dissolve 1.00 g of primary standard grade $CaCO_3$ in dilute HCl. Evaporate to dryness and dilute to 1000 ml in a volumetric flask. (1 ml = 1.00 mg $CaCO_3$.)
Standard EDTA Solution. Dissolve 3.72 g of disodium dihydrogen ethylenediaminetetraacetate dihydrate in 800 ml of water and add 21.5 ml of 1N NaOH. Store in Pyrex or polyethylene. Check the titer of the solution once a week. (1 ml is equivalent to 1.00 mg $CaCO_3$.)
Procedure: Transfer a 100-ml sample of water to a Coors No. 3A porcelain casserole.

Add 1.0 ml of buffer solution and 0.2 g of solid Eriochrome Black T indicator, stirring between additions.

Titrate slowly with standard EDTA solution until the color changes from wine-red to clear blue and record the buret reading.

$$\text{Total hardness, ppm } CaCO_3 = \text{ml titration} \times 10$$

Calcium Hardness – CaH

Reagents: Standard EDTA Solution. Prepare as in the above procedure.
Sodium Hydroxide – 1N Approximately. Dissolve 40 g of NaOH in 1000 ml of water.
Calcium Indicator. Mix 0.2 g of ammonium purpurate (Eastman No. 6733) with 100 g of NaCl, grind to 40-50 mesh, and store in a dark bottle.
Procedure: Transfer a 100-ml sample of water to a Coors No. 3A porcelain casserole.

Add 4.0 ml of 1N NaOH and 0.2 g of solid calcium indicator, stirring between additions.

Titrate slowly with standard EDTA solution until the color changes from salmon-pink to orchid, and record the buret reading.

$$\text{Calcium hardness, ppm } CaCO_3 = \text{ml titration} \times 10$$

f. Determination of Hydrazine

Hydrazine readily condenses with p-dimethylaminobenzaldehyde (I) to form the hydrazone (II), which itself condenses with a second molecule of (I) to form the corresponding bright yellow azine (III). This reaction is the basis of a colorimetric method devised by Watt and Chrisp[13] for determining hydrazine in water.

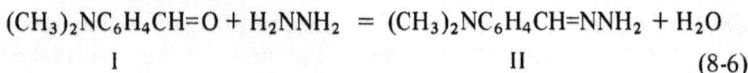

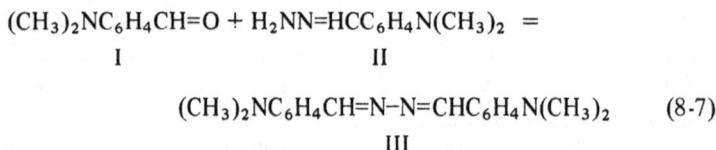

$$(CH_3)_2NC_6H_4CH=O + H_2NNH_2 \; = \; (CH_3)_2NC_6H_4CH=NNH_2 + H_2O$$

$$\text{I} \hspace{6cm} \text{II} \hspace{3cm} (8\text{-}6)$$

$$(CH_3)_2NC_6H_4CH=O + H_2NN=HCC_6H_4N(CH_3)_2 \; =$$

$$\text{I} \hspace{4cm} \text{II}$$

$$(CH_3)_2NC_6H_4CH=N-N=CHC_6H_4N(CH_3)_2 \hspace{1cm} (8\text{-}7)$$

$$\text{III}$$

The control point for hydrazine dosage, and thus the sampling point for measuring hydrazine, should be at the economizer or in the feed line after the last stage of feed water preheat. On ships, hydrazine is pumped continuously into the crossover line between the high-pressure and low-pressure turbines when steaming at sea, and into the storage section of the deaerator when in port. Dosage should be adjusted daily so that residual hydrazine in the feed water is at the lowest value detectable—preferably near the low end of the range 0.01–0.05 ppm—to avoid the formation of excessive concentrations of ammonia in the steam condensate system. Samples of blowdown water may be checked occasionally to verify that there is a hydrazine residual in the boiler water. The concentration here should be higher than in the feed water, but the actual value depends on the pressure of the boiler (see Table 5.7).

To obtain reliable hydrazine results, the following precautions must be observed when sampling.

1. Do not collect a sample until the flowing stream is clear without particles of rust.
2. If the sample stream does not clear, install a line filter before collecting a sample for the hydrazine determination.
3. Make sure that the sampling valve is at the outlet of the sampling line rather than at the inlet.

Hydrazine in Feed Water

Reagents: *p-Dimethylaminobenzaldehyde Solution.* Dissolve 4.0 g of *p*-dimethylaminobenzaldehyde (Eastman No. 95) in 200 ml of methanol and 20 ml of HCl. Store in a dark bottle.

Hydrazine Standard Solution. Dissolve 0.328 g of $N_2H_4 \cdot 2HCl$ (Eastman No. 1117) in 100 ml of water and 10 ml of HCl, transfer to a 1000-ml volumetric flask, dilute to volume, and mix thoroughly. (1 ml = 0.10 mg N_2H_2.)

Dilute Hydrazine Standard Solution. Transfer 10.0 ml of hydrazine standard solution to a 1000-ml volumetric flask containing 500 ml of water and 10 ml of HCl. Dilute to volume and mix. (1 ml = 0.001 mg N_2H_4 = 1 μg N_2H_4.)

Dilute Hydrochloric Acid. Dilute 10 ml of HCl to 100 ml in a 100-ml glass-stoppered mixing cylinder, and mix thoroughly.

Standardization: Transfer 2.0, 5.0, 8.0, and 10.0 ml of dilute hydrazine standard solution, corresponding to the same number of micrograms of hydrazine, to a series of 50-ml glass-stoppered cylinders, each containing 5.0 ml of dilute HCl, and include a blank.

Bring the total volume in each cylinder to 50 ml, and mix. To each cylinder add 10.0 ml of *p*-dimethylaminobenzaldehyde solution, stopper, and mix thoroughly.

After 10 min measure the transmittance of each solution at 458 nm, using the blank solution to set zero.

Plot the transmittances against the corresponding μg of N_2H_4 on semilogarithmic graph paper.

Procedure: Collect a clear 45-ml sample of water in a 50-ml glass-stoppered mixing cylinder containing 5.0 ml of dilute hydrochloric acid.

Add 10.0 ml of *p*-dimethylaminobenzaldehyde solution to the contents of the mixing cylinder. Stopper, mix thoroughly, and after 10 min, measure the transmittance of the solution at 458 nm against a reagent blank.

Read the micrograms of hydrazine from the prepared calibration curve.

$$\text{ppm hydrazine, } N_2H_4 = \mu g\ N_2H_4 / 45$$

g. Insoluble Contamination in Condensate

In order to measure the effectiveness of corrosion inhibitors in condensate systems, the concentration of iron and copper is often determined in condensate, although this is a complicated procedure when dealing with the small concentrations (5-50 ppb) that are present when corrosion is under control. A more useful method is a visual examination of the stain left on a Millipore filter pad after filtering a specified volume of condensate. This is a practical, empirical method for evaluating the performance of filming or neutralizing amines that is quite satisfactory for use in a boiler plant. As there is no consistent correlation between the degree of corrosion protection and the concentration of residual amine in condensate, there is no point in testing for amine. In the filter test, however, products of corrosion are concentrated to form a visible stain, the intensity of which is directly proportional to the extent of the corrosion. Before collecting a sample for filtration, the condensate stream should be allowed to flow continuously, without disturbing the sampling valve or jarring the cooling coil, for at least 30 min.

Millipore Filter Test for Contamination

Equipment: Millipore All-Glass Filter Apparatus, 47 mm. Catalog No. XX 15 047 00, Millipore Corporation, Bedford, Massachusetts 01730. *Millipore Filters.* Type HAWP, 0.45 μm. Catalog No. HAWP 047 00.

Procedure: If possible, collect the flowing condensate directly in the funnel of the Millipore filter apparatus, which has been fitted with a 47-mm, 0.45-μm filter.

After one liter of filtered condensate has been collected in the filter flask, discard the water remaining in the funnel, and dismantle the assembly.

Examine the filter, noting the intensity of the stain; the darker the stain, the more severe the corrosion. The procedure can be made semiquantitative by dissolving the material on the filter and determining iron and copper.

h. The Measurement of pH

The adjustment of pH and the chemistry of carbonic acid and its salts play predominant roles in conditioning water for boilers. Fig. 8.1 illustrates some of the relationships among pH and the usual forms of alkalinity

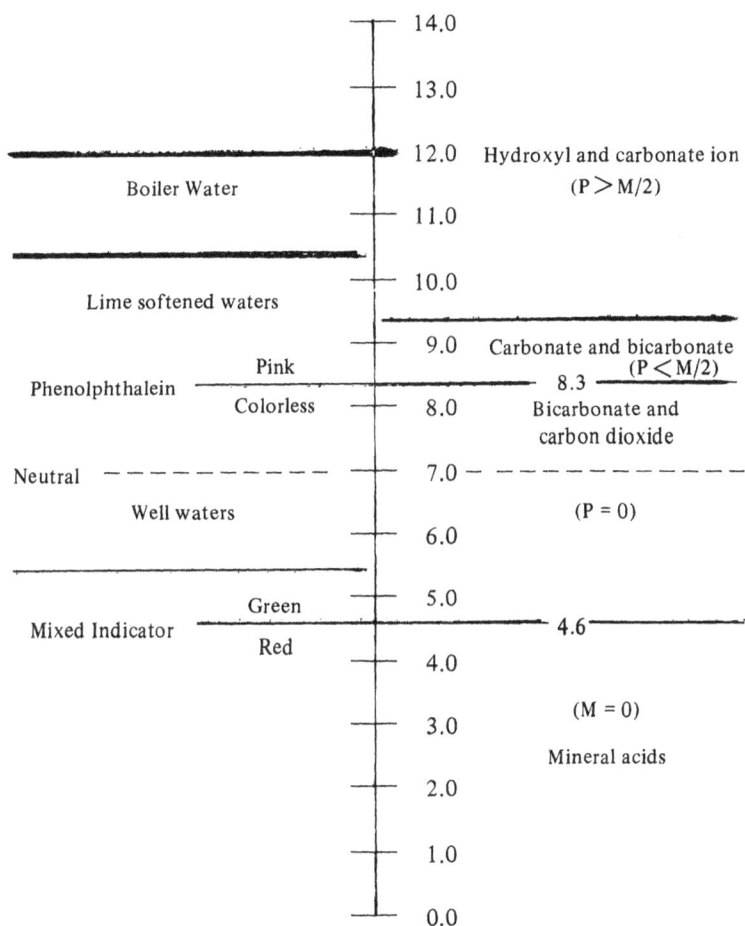

Fig. 8.1   pH in water treatment.

in boilers, make up waters, condensates, and raw waters. In order to control boilers and equipment used for the external treatment of make up water, it is essential that reliable pH measurements be made. To do this, an indicating glass electrode, sensitive to the concentration of hydrogen ion, is combined with a reference electrode having an opposed potential. The difference in potential, which is proportional to pH, is measured with a high-impedance voltmeter capable of detecting a very small flow of current. Circuits are so adjusted that the opposed potentials are equal when the two electrodes are immersed in a solution buffered at pH 7. In acidic solutions (pH < 7) the potential of the glass electrode exceeds that of the reference; in alkaline solutions (pH > 7) the potential of the glass electrode is lower than that of the reference electrode. When not in use, the electrodes should be immersed in a solution buffered at pH 4. With time, the rate of response of the glass electrode becomes slow. Sensitivity can be increased by dipping the electrode alternately in a tenth normal hydrochloric acid and a tenth normal sodium hydroxide, 1 min in each solution, twice. Failure to allow sufficient time for equilibrium to be established frequently leads to faulty results.

It is difficult to obtain a reliable pH measurement in demineralized water or deaerated condensate as these solutions are unbuffered and contain few ions to complete the circuit between the electrodes. A pinch of sodium chloride usually stabilizes the solution and eliminates drift in the pH meter. Wescott[14] has published an exemplary treatment of the practical aspects of making pH measurements. So many different pH meters are manufactured, and their use is so common, that no attempt to write a detailed procedure for measuring pH will be made. The meter is standardized, however, by immersing the electrodes in a standard buffer solution, and then adjusting the meter to indicate the defined value, pH_s. The electrodes are rinsed thoroughly, then immersed in the unknown solution, the pH of which is indicated by the meter.

i. Determination of Phosphate

As already explained, ortho-phosphate is used in the treatment of boiler water for precipitating alkaline earth elements in a soft dispersed form to prevent the formation of scales on boiler tubes. The following procedure makes use of the molybdenum blue color produced by the selective reduction of molybdiphosphoric acid $[H_3P(Mo_3O_{10})_4]$ resulting from the condensation of ortho-phosphate and molybdate ions in weak acid solution. As only two reagents are used, and the reaction is fairly specific as applied to most water samples, the procedure is recommended for the routine determination of ortho-phosphate in boiler water.

Greenberg, et al.[15] have called attention to the interference by nitrite in this method. As nitrite is a decomposition product of sodium nitrate, used to inhibit stress corrosion cracking in some boilers, their recommendation of adding sulfamic acid to the molybdate reagent has been followed. The reaction between nitrite (or nitrous acid) and sulfamic acid is as follows:

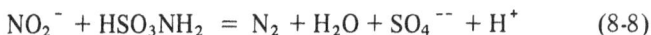

$$NO_2^- + HSO_3NH_2 = N_2 + H_2O + SO_4^{--} + H^+ \qquad (8\text{-}8)$$

Ortho-Phosphate in Boiler Water

Reagents: Molybdate Solution. Dissolve 37.5 g of $(NH_4)_6Mo_7O_{24}\cdot4H_2O$ in about 600 ml of water. Slowly add 300 ml of H_2SO_4, and cool. To the solution add 10 g of HSO_3NH_2, stir to dissolve, and dilute to 1000 ml.
Sulfite Reducing Reagent. Dissolve 80 g of $Na_2S_2O_5$ (sodium metabisulfite) in 800 ml of water. Dissolve 7 g of Na_2SO_3 and 1.5 g of 1-amino-2-naphthol-4-sulfonic acid (Eastman No. 360) in 100 ml of water; stir until dissolved, then combine with the first solution and dilute to 1000 ml with water. (Discard this reagent if it gives a green instead of a blue color in the procedure below.)
Standard Phosphate Solution. Dissolve 1.433 g of KH_2PO_4 in water and dilute to 1000 ml. (1 ml = 1.00 mg PO_4^{---}.)
Standardization: Dilute 25.0 ml of phosphate standard solution to 100 ml to produce a solution containing 0.25 mg PO_4^{---}/ml. Transfer 2.0, 4.0, 6.0, 8.0, and 10.0 ml of the dilute solution, corresponding to 0.5, 1.0, 1.5, 2.0, and 2.5 mg of PO_4^{---}, to a series of 100-ml volumetric flasks, and include a blank.

Bring the total volume of each flask to about 80 ml, then while swirling, add 5 ml of molybdate reagent, followed by 5 ml of sulfite reducing reagent. Dilute to volume and mix thoroughly.

Measure the transmittances of the solutions at 650 nm between 7 and 10 min after adding the second reagent. Plot the results against the corresponding mg of PO_4^{---} on semilogarithmic graph paper.
Procedure: Filter the sample if it is not clear, transfer a 50-ml portion to a 100-ml volumetric flask, and dilute to about 80 ml.

Add, while swirling, 5 ml of molybdate reagent, followed by 5 ml of sulfite reducing reagent. Dilute to the mark and mix thoroughly.

Measure the transmittance of the solution at 650 nm, 7-10 min after adding the second reagent, and determine the mg of PO_4^{---} from the prepared calibration curve.

$$\text{ppm PO}_4^{---} = \text{mg PO}_4^{---} \times 20$$

j. Determination of Silica

The method prescribed here is generally applicable to raw, feed, and boiler waters, for determining soluble silicates. As pointed out by Dwyer and Frith,[16] however, colloidal silicic acid (H_4SiO_4), silica, and some polymers ($mSiO_2 \cdot nH_2O$) are not detected by this procedure. For this reason, blowdown ratios calculated from silica values may be erroneously high, because part of the colloidal silica may be unreactive in the feed water, but is hydrolyzed in the boiler to a reactive form. Thus, the concentration determined in the feed water will be too low, whereas that in the boiler water will be correct. This phenomenon is of slight importance in boilers operated below 1500 psi, but can be significant at higher pressures, especially in supercritical boilers. The magnitude of the errors in an extreme case are shown in Table 8.2.

TABLE 8.2

Silica Forms in Demineralized Water

| Sample | ppm SiO_2 | | | Percent unreactive |
	Reactive	Unreactive	Total	
1	0.019	0.044	0.063	70
2	0.018	0.103	0.121	85
3	0.008	0.058	0.066	88
4	0.008	0.039	0.047	83

Schwartz[17] has described the determination of soluble silica in water using the yellow color of the complex silico-12-molybdate, $SiMo_{12}O_{40}^{-4}$, formed by reacting silicate with ammonium molybdate solution. Phosphate forms a similar complex phosphomolybdic acid, but it is destroyed by the addition of sodium citrate, which has the additional benefit of eliminating a disturbing effect of tannins. There is no significant interference by normal amounts of phosphate, tannins (including pyrogallic acid, tannic acid, and pyrocatechol), iron, chloride, alkalinity, sulfate, or hardness. Also, if samples are analyzed immediately, glass containers may be used, but the reagents should be stored in polyethylene.

Silica in Feed and Boiler Water

Reagents: Ammonium Molybdate Solution. Dissolve 75 g of $(NH_4)_6Mo_7$-$O_{24} \cdot 4H_2O$ in 800 ml of water, add 60 ml of H_2SO_4, and dilute to 1000 ml.
Sodium Citrate Solution. Dissolve 430 g of $Na_3C_6H_5O_7 \cdot 2H_2O$ in water, and dilute to 1000 ml.
Sulfite Reducing Reagent. Prepare as described in the procedure for phosphate.
Standard Silica Solution. Dissolve 4.8 g of $Na_2SiO_3 \cdot 9H_2O$ in water and dilute to 1000 ml. Standardize gravimetrically and adjust so that 1 ml = 1.00 mg SiO_2.
Standardization: Transfer 25.0 ml of the standard silica solution to a 100-ml volumetric flask and dilute to the mark with water. (1 ml = 0.25 mg SiO_2.)

Transfer 1.0, 2.0, 3.0, 4.0, and 5.0 ml of the dilute standard to a series of 100-ml volumetric flasks and include a blank.

Dilute the contents of each flask to about 50 ml, then add to each, 5 ml of ammonium molybdate solution, and let stand 5 min.

To each flask add 20 ml of sodium citrate solution while swirling, followed immediately by 5 ml of reducing reagent.

Dilute to the mark with water, mix, let stand 7-10 min, then determine the transmittance of the solutions at 650 nm.

Plot the transmittances observed against the number of milligrams of silica on semilogarithmic graph paper.

Procedure: Measure a suitable sample and transfer it to a 100-ml volumetric flask, diluting to 50 ml, if necessary. For water containing 0-30 ppm SiO_2 use 50 ml; 10-60 ppm use 25 ml; 50-300 ppm use 5 ml of sample.

Add 5 ml of molybdate solution and let stand 5 min.

While swirling, add 20 ml of sodium citrate solution, followed by 5 ml of reducing reagent.

Dilute to the mark with water, mix, let stand 7-10 min, then determine the transmittance of the solution at 650 nm.

From the prepared calibration curve read the corresponding mg of SiO_2, correct for a reagent blank, if necessary, and calculate the ppm SiO_2 in the sample.

$$ppm\ SiO_2 = mg\ SiO_2 \times 1000/ml\ sample$$

k. Determination of Sulfite

Sulfite is oxidized to sulfate by iodine in acidic solution as follows:

$$SO_3^{--} + I_3^- + H_2O = SO_4^{--} + 3I^- + 2H^+ \qquad (8\text{-}9)$$

If the sample is acidified and titrated directly, however, low results are obtained because sulfurous acid is oxidized in part by air, and to a lesser extent, because of the volatilization of sulfur dioxide from solutions of strong acids. Accurate results are obtained by titrating an acidified solution with standard iodate-iodide solution. The titration can be thought of as occurring in two successive steps.

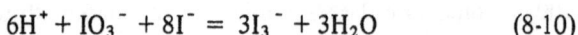

$$6H^+ + IO_3^- + 8I^- = 3I_3^- + 3H_2O \qquad (8\text{-}10)$$

follwed by the reaction in Eq. (8-9), but the net reaction, which is essentially instantaneous, is

$$3SO_3^{--} + IO_3^- = 3SO_4^{--} + I^- \qquad (8\text{-}11)$$

Sulfite in Boiler Water

Reagents: Standard Iodate-Iodide Solution. (0.0125 N; 1 ml = 0.50 mg SO_3^{--}) Dissolve 0.31 g of $NaHCO_3$, 4.35 g of KI, and 0.446 g of KIO_3 in water, dilute to 1000 ml, and mix.

Thyodene Indicator. Catalog No. T-138, Fisher Scientific Company, Fair Lawn, New Jersey 07410.

Procedure: Transfer 0.5 ml of HCl to a Coors No. 3A porcelain casserole.

Transfer a 100-ml freshly drawn sample of boiler water to the casserole, add 0.5 g of Thyodene indicator, and stir gently until dissolved.

Titrate with standard iodate-iodide solution to a pale blue end point, and record the titration.

$$ppm\ SO_3^{--} = ml\ iodate\text{-}iodide \times 5$$

8.2 ANALYSIS OF CHEMICAL CLEANING SOLUTIONS

In this section methods are given for checking the strengths of the cleaning solutions described in Section 7.7a, b, and c. In all of the following procedures, it is to be understood that samples of chemical cleaning solutions should be filtered before measuring a portion for analysis.

a. Available Hydrochloric Acid

Hydrochloric acid titrates as a monobasic acid to phenolphthalein indicator with standard alkali solution. Account must be taken, though, of the presence of hydrolyzable metallic ions, notably ferric ion, which behaves as an acid toward phenolphthalein. Ferric ion can be rendered innocuous by adding potassium fluoride, which forms the very stable hexafluoroferric ion.

$$Fe^{+++} + 6F^- = FeF_6^{---} \qquad (8\text{-}12)$$

Reagent: *Potassium Fluoride Solution.* Dissolve 30 g of $KF \cdot 2H_2O$ in 100 ml of water and store in a polyethylene bottle.
Procedure: Transfer by means of a pipet a 10-ml portion of filtered sample to a 250-ml Erlenmeyer flask containing about 100 ml of water.
    Add 20 ml of 30 percent KF solution, swirl, and titrate to phenolphthalein with standard 0.5 N NaOH. Record the titration.

$$\text{percent HCl} = (\text{titr.} \times N \times 36.5) / (\text{ml sample} \times 10)$$

(As approximate concentrations only are needed, there is no necessity for using specific gravity of the sample in the calculation.)

b. Caustic and Soda Ash in Boil-Out Solutions

As discussed in Section 7.7a, alkaline boil-outs are done in new boilers to remove oil, grease, and other debris introduced during manufacture and erection of boilers. These solutions are composed of a mixture of sodium hydroxide and sodium carbonate; these two components can be estimated on the basis of the relationships shown in Table 5.4.

Reagent: Mixed Indicator. Dissolve 0.02 g of Methyl Red (Eastman No. 431) and 0.1 g of Bromcresol Green (Eastman No. 1782) in 100 ml of Formula 30 alcohol.

Procedure: Transfer a 10-ml portion of the filtered sample to a 250-ml Erlenmeyer flask containing about 100 ml of water.

Titrate to phenolphthalein indicator with standard 0.5 N HCl, and record the titration as P ml.

Add mixed indicator and continue the titration to the red end point, recording the total titration as M ml.

$$\text{Percent NaOH} = [(2P - M) \times N \times 40] / [\text{ml sample} \times 10]$$

$$\text{Percent Na}_2\text{CO}_3 = [2(M - P) \times N \times 53] / [\text{ml sample} \times 10]$$

c. Estimation of Residual Chelant

This and the next two procedures are intended for following the course of alkaline cleaning with ammoniacal EDTA, as described on pp 233-236. During the first stage of the cleaning procedure, residual chelant is monitored to make sure that an excess of EDTA is present. The cleaning solution in the boiler is then allowed to cool, after which air is blown into the boiler, and the dissolution of copper commences. During the latter stage, residual chelant is also monitored.

Residual chelant is estimated by titrating with a standard solution of strontium chloride in the presence of oxalate at a pH of 9.0-9.5. Excess chelant sequesters strontium ion until an excess is present, when the solution becomes turbid with a precipitate of strontium oxalate.

Reagents: Ammonium Oxalate – Saturated Solution. Dissolve 60 g of $(NH_4)_2C_2O_4 \cdot H_2O$ in 1000 ml of water.

Strontium Chloride Standard Solution. (0.50 M.) Dissolve 133 g of $SrCl_2 \cdot 6H_2O$ in 800 ml of water, and dilute to volume in a 1000-ml volumetric flask.

Procedure: Transfer a 25-ml sample of cleaning solution to a 250-ml Erlenmeyer flask containing 100 ml of water, and add 5 ml of saturated ammonium oxalate solution.

Titrate the solution with 0.5 M $SrCl_2$ solution to the first permanent turbidity.

Check the pH with test paper and if the pH < 9.0, add NH_4OH until it is in the range of 9.0-9.5.

If the precipitate dissolves, continue titrating with $SrCl_2$ until a permanent turbidity is formed, then record the titration.

Percent Na_4EDTA = (Titr. $\times$ N $\times$ 380)/(ml sample $\times$ 10)

d. Copper in EDTA Cleaning Solutions

A sample of cleaning solution is boiled with persulfate to destroy EDTA and to ensure that all iron is in the ferric state. The acidity of the solution is adjusted with acetic acid and fluoride is added to inactivate ferric ion. Potassium iodide is added, after which the liberated iodine is titrated to the starch end point with standard thiosulfate solution.

$$2Cu^{++} + 7I^- = 2CuI_2^- + I_3^- \qquad (8\text{-}13)$$

Reagents: Thyodene Indicator. Catalog No. T-138, Fisher Scientific Company, Fair Lawn, New Jersey 07410.

Procedure: Transfer a 5.0-ml sample of cleaning solution to a 250-ml beaker containing 100 ml of water, add 2 g of $(NH_4)_2S_2O_8$, heat to boiling, and boil 10 min, then cool to room temperature.

Add NH_4OH dropwise until a brown precipitate of $Fe(OH)_3$ forms, then add 10 ml of acetic acid.

Add 1 g of NH_4HF_2, stir until dissolved, then add 3 g of KI and 0.5 g of Thyodene indicator.

Titrate with standard 0.1 N $Na_2S_2O_3$ to the disappearance of the blue color, and record the titration. (This titration is needed if iron is to be determined, as described in the following procedure.)

Percent copper = (ml $S_2O_3^{--}$ $\times$ N $\times$ 63.6)/(ml sample $\times$ 10)

e. Iron in EDTA Cleaning Solutions

In this procedure both copper and iron oxidize iodide to iodine so the total titration obtained here must be corrected by subtracting that obtained in the foregoing procedure for copper.

Procedure: Transfer a 5.0-ml sample of cleaning solution to a 250-ml beaker containing 100 ml of water, add 2 g of $(NH_4)_2S_2O_8$, heat to boiling and boil 10 min, then cool to room temperature.

Add NH_4OH dropwise until a brown precipitate of $Fe(OH)_3$ forms, then add 10 ml of acetic acid.

Add 3 g of KI and 0.5 g of Thyodene indicator, then titrate with 0.1 N $Na_2S_2O_3$ to the disappearance of the blue color, and record the titration.

Percent iron = [(ml total titr. − ml copper titr.) × N × 56] /(ml sample × 10)

8.3   EVALUATION OF ION EXCHANGE RESINS

When ion exchange resins begin to show signs of decreased capacity it is the custom in many water-treating plants to submit samples of resin to a laboratory for a determination of total capacity. It is relatively rare, however, for a cation exchange resin to lose significant exchange capacity and in most instances any apparent loss is caused by improper operating procedures, the most common of which are the following:

1.   Attrition of the resin beads.
2.   Loss of resin through excessive backwashing.
3.   Fouled resin as the result of insufficient backwashing.
4.   Excessive dilution of regenerant in the water eductor.
5.   Improper flow rate of regenerant.
6.   Poor internal distribution of regenerant.
7.   Channeling in the resin bed.

Anion exchangers are much more susceptible to damage and loss of exchange capacity than are cation exchangers. Among the causes are:

1.   Thermal deterioration of strong base resins.
2.   Damage by oxidizing agents.
3.   Contamination by degradation products of cation exchange resins.
4.   Poisoning by organic material in raw water.

As the most prevalent causes of the loss of capacity of strongly basic anion exchange resins are related to the transformation of the active quaternary ammonium group to a tertiary amine, it is usually sufficient to measure the proportion of the two forms. A procedure for doing this is given in Section 8.3a. As inefficient performance of cation exchange resins is often attributable to deficiencies in the regenerating process, these can usually be disclosed by plotting an elution curve, as described in Section 8.3b. The determination of total exchange capacity and other analyses have been covered in another publication.[18]

a.   Salt-Splitting Capacity of Strongly Basic Resins

Strong-base anion exchange resins are composed of quaternary ammonium groups attached to a polystyrene structure. The four substituents on the nitrogen atom are three methyl groups and a polymeric benzyl group, which in the hydroxide form, can exchange hydroxyl ions for other anions.

$$
R\left[CH_2-\underset{\underset{CH_3}{|}}{\overset{\overset{CH_3}{|}}{N}}-CH_3\right]^+ OH^- + Cl^- = R\left[CH_2-\underset{\underset{CH_3}{|}}{\overset{\overset{CH_3}{|}}{N}}-CH_3\right]^+ Cl^- + OH^-
$$

(8-14)

Hoffman[19] discovered a century ago that quaternary ammonium hydroxides undergo thermal degradation to tertiary amine and an alkene or methanol, depending upon configuration. Thus, if the strongly basic anion exchange resins are overheated, the following reaction occurs.

$$
R\left[CH_2-\underset{\underset{CH_3}{|}}{\overset{\overset{CH_3}{|}}{N}}-CH_3\right]^+ OH^- = R\left[CH_2-\underset{\underset{CH_3}{|}}{\overset{\overset{CH_3}{|}}{N}}\right] + CH_3OH
$$

(8-15)

$$\qquad\qquad I \qquad\qquad\qquad\qquad II$$

In this way, the quaternary ammonium hydroxide (I) is converted to a tertiary amine (II), which cannot exchange anions of neutral salts. This deficiency is revealed by determining a resin's salt-splitting capacity, as described in the following procedure. A measured volume of resin in the hydroxide form is exhausted with dilute sodium chloride, and its strong-base capacity is measured by titrating the effluent with standard acid.

Reagents: Sodium Hydroxide, 2 N, approximately. Dissolve 80 g of NaOH in 1000 ml of water.

Sodium Sulfate Solution, 0.5 N, approximately. Dissolve 36 g of Na_2SO_4 in 1000 ml of water.

Apparatus: Glass Column. 2.5 cm in diameter, about 100 cm long, fitted with a stopcock and containing a fritted glass plate just above the stopcock.

Procedure: Measure 50 ml of wet resin in a graduated cylinder and transfer it quantitatively to the glass column. Allow the resin to settle, then mark the column at the surface of the resin.

Backwash for a few minutes to eliminate fines and flocculent material, then regenerate the resin by passing 1000 ml of 2N NaOH through the column at 6-8 ml/min.

Rinse with demineralized water until the effluent is neutral to phenolphthalein (pH < 8.3), then exhuast the resin by passing 1000 ml of 0.5 N Na_2SO_4 through the column at 12-14 ml/min, receiving the effluent in a 2-liter beaker.

Rinse the column with 500 ml of demineralized water, collecting the rinse water with the alkaline effluent in the same 2-liter beaker.

Titrate the combined effluent with a standard 1N HCl to the phenolphthalein end point, and record the titration.

Using a ruler, measure the height of the resin column and calculate its volume by proportion from the original mark.

$$\text{Salt-splitting capacity, meq/ml} = (\text{ml titr.} \times \text{N HCl}) / (\text{ml resin})$$

(New strongly basic anion exchange resin has a capacity of 1.3-1.4 meq/ml.)

b. Elution Pattern of Strong-Acid Resins

When regenerating a cation exchanger with salt solution, the regenerant should flow at the rate of 1 gal/min/ft^3 of resin. The optimum concentration is 10 percent by weight of NaCl per cubic foot of resin. A brine tank of suitable capacity is filled with saturated salt solution (26.3 percent NaCl @ 60 F), which is then discharged into the softener by a water eductor sized to dilute the brine to 10 percent NaCl in the process. As an example, a softener containing 100 ft^3 of resin requires 1400 lb of NaCl, or $1400/0.263 = 5323$ lb of saturated solution. The specific gravity of a saturated solution is 1.203 @ 60 F, equivalent to $1.203 \times 8.34 = 10.0$ lb/gal. Therefore, $5323/10.0 = 532$ gal of saturated brine are needed to regenerate 100 ft^3 of resin. This saturated brine is to be diluted to 10 percent by water from the eductor, at which concentration the specific gravity is 1.074, or 8.96 lb/gal. At 10 percent concentration the weight of solution required is $1400/0.10 = 14,000$ lb, so the amount of water through the softener must be $14,000/8.96 = 1563$ gal. As the desired rate is 1 gpm/ft^3, the time to regenerate is $1563/100 = 15.6$ min.

To the times recorded in the last column of Table 8.3, however, the time required to empty the voids in the resin bed must be added, which is assumed to contain 40 percent voids. As 1 ft^3 = 7.5 gal, the volume to be

TABLE 8.3

Properties of Brine Solutions

Percent NaCl	Percent saturation	Sp. Gr. @ 60 F	lb/gal	Gallons per regeneration	Minutes per regeneration
26.3	100	1.203	10.00	532	5.3
10.0	38	1.074	8.96	1563	15.6
8.0	30	1.059	8.83	1981	19.8

displaced at the rate of 100 gpm is $7.5 \times 100 \times 0.4 = 300$ gal, which at 100 gpm takes 3 min. Taking account of dilution at the leading and trailing brine-fresh water interfaces, the length of time that the concentration of salt in water from the softener exceeds 8 percent (30 percent saturation) should be 25–30 min.

During the application of brine the concentration of salt rises to some maximum value, then falls off. By measuring the concentration of salt in water from the drain line at the softener at regular intervals with a salinometer (salometer), the efficiency of the regeneration can be checked. Poor regeneration can result from a number of causes:

1. Insufficient brine solution.
2. Brine solution too dilute.
3. Partially plugged line from brine tank to eductor.
4. Distribution nozzles inside softener blocked.
5. Channeling through the resin bed.
6. Improper flow rates of brine or rinse water.
7. Fouling in the resin bed.
8. Insufficient backwash.

Equipment: Salinometer. Special hydrometer with scale graduated from 0–100 percent saturation, or 0–26.5 percent NaCl.

Hydrometer Jar. A 1000-ml cylinder is satisfactory.

Procedure: When backwashing is complete and the softener is ready to be regenerated, take a sample of water from the drain and note the time when the multiport valve switches to the "brine" position, then take samples every 5 min until the concentration of brine falls to 5 percent saturation or 1.5 percent NaCl.

Plot the times as abscissa points against salinometer readings as ordinates on rectangular coordinate paper.

Draw a horizontal line through the plot at 30 percent saturation (8 percent NaCl), and note the number of minutes that the salinometer readings are on or above the horizontal line. If the regeneration is satisfactory, this time should be 25–30 min; if it is less, it is likely that one or more of the circumstances listed above is the cause.

8.4 EQUIVALENT WEIGHT OF AMINE FORMULATIONS

In order to calculate the dosage of neutralizing amine needed to react with carbon dioxide in condensate, to determine the cost of treatment, and to evaluate the relative costs of treating with different proprietary formulations, it is necessary to find the equivalent weights of the formulations. Furthermore, if the intention is to neutralize the condensate to a pH of 8, or thereabout, phenolphthalein should be used as the indicator in determining the equivalent weight, rather than bromcresol green, which measures both weak and strongly basic amines. Table 8.4 illustrates these points.

TABLE 8.4

Basic Strength of Neutralizing Amines

	Milliequivalents per gram		Grams per equivalent	
Formulation	Basicity to pH 8.3	Basicity to pH 5.0	Equiv. weight to pH 8.3	Equiv. weight to pH 5.0
A	3.48	3.92	287	255
B	6.63	12.80	151	78
C	7.80	9.83	128	102

The formulation of choice is that with the highest basicity and lowest equivalent weight measured at pH 8.3, namely, C. Using basicity at pH 5.0 (bromcresol green), B would appear to be superior to the other two formulations, but this is not the proper interpretation, as B contains a high proportion of weak bases that cannot raise the pH to the desired level. In evaluating the effectiveness of these formulations, unit prices must also be compared. To do this, multiply the equivalent weight at pH 8.3 of each formulation by its cost in dollars per pound. This gives the cost of neutralizing 44 lb of carbon dioxide with each inhibitor; that requiring the least number of dollars, of course, is the most economical formulation to use, regardless of basicities.

Procedure: Accurately weigh approximately 1 g of neutralizing amine solution, and transfer it to a 250-ml Erlenmeyer flask containing 100 ml of water.

Titrate with standard 0.5 N HCl to the phenolphthalein end point, if the formulation is to be used to raise the pH of condensate to pH 8.3, or to bromcresol green, if total amines are to be measured, and record the titration.

$$\text{Basicity, meq/g} \ = \ \text{titr.} \times \text{N HCl/g sample}$$

$$\text{Equivalent weight, g/eq} \ = \ 1000/\text{basicity}$$

REFERENCES

(1) American Public Health Association. 1978. *Standard methods for the examination of water and wastewater. 14th ed.* New York.

(2) American Society for Testing Materials. *1977 Annual Book of ASTM Standards.* Part 31. Water.

(3) Mohr, Fr. 1856. Neue massanalytische bestimmung des chlors in verbindungen. *Ann.* 97:335.

(4) Sheen, R. T. and Kahler, H. L. 1938. Effect of ions on Mohr method for chloride determination. *Ind. Eng. Chem.; Anal. Ed.* 10:628.

(5) Maguire, J. J. and Polsky, J. W. May, 1947. Simplified plant control test for boiler water dissolved solids. *Combustion* 18 (11):35.

(6) Schwarzenbach, G., Biedermann, W., and Bangerter, F. 1946. Complexons VI. new simple titrating methods for determining the hardness of water. *Helv. Chim. Acta* 29:811.

(7) Schwarzenbach, G. and Biedermann, W. 1948. Alkaline earth complexes of o,o-dihydroxy azo dyes. *Helv. Chim. Acta* 31: 678.

(8) Diehl, H., Goetz, C. A., and Hach, C. C. 1950. The versenate titration for total hardness. *J. Amer. Water Works Assn.* 42: 40.

(9) Betz, L. D. and Noll, C. A. 1950. Total hardness determination by direct colorimetric titration. *J. Amer. Water Works Assn.* 42:49.

(10) Marcy, V. M. Jan., 1950. New water hardness test is faster and gives more accurate results. *Power* 94 (1):105.

(11) American Society for Testing Materials. *1977 Annual Book of ASTM Standards.* Part 31. Water, Method D1126-67. Standard test methods for hardness in water.

(12) Connors, J. J. 1950. Advances in chemical and colorimetric methods. *J. Amer. Water Works Assn.* 42:33.

(13) Watt, G. W. and Chrisp, J. D. 1952. Spectrophotometric method for the determination of hydrazine. *Anal. Chem.* 24: 2006.

(14) Wescott, C. C. 1978. *pH measurements.* New York: Academic Press.

(15) Greenberg, A. E., Weinberger, L. W., and Sawyer, C. N. 1950. Control of nitrite interference in colorimetric determination of phosphorus. *Anal. Chem.* 22:499.

(16) Dwyer, J. L. and Frith, C. F. 1967. New analytical techniques for determination of colloidal contaminants in high purity steam generating systems. *Proc. Amer. Power Conf.* 29:778.
(17) Schwartz, M. C. 1942. Photometric determination of silica in the presence of phosphates. *Ind. Chem.; Anal. Ed.* 14:893.
(18) McCoy, J. W. 1969. *Chemical analysis of industrial water.* New York: Chemical Publishing.
(19) Hofmann, A. W. 1881. Einwirkung der wärme auf die ammoniumbasen. *Ber.* 14:659.

Appendix 1.

Glossary of Terms

Absolute pressure. Gauge pressure plus atmospheric pressure (14.7 psi).

Anion exchange. Replacement of negative ions of salts in solution by hydroxyl ions by passage through a basic resin bed.

Anode. The positive electrode of an electrochemical cell where electrons are donated and oxidation occurs.

Azeotrope. One of two or more compounds that form mixtures having a constant boiling point.

Backwash. Passage of water upward through a filter or an ion exchange bed to remove floc and accomplish hydraulic grading.

Base-loaded boiler. A steam generator operated continuously at a constant output of steam.

Blowdown. A portion of recirculating water that is continuously or intermittently removed from a boiler to limit the concentration of soluble or insoluble material.

Blowdown ratio. The concentration of dissolved solids in boiler water divided by that in feed water. Analogous to cycles of concentration in recirculating cooling systems.

Boiler code. A set of rules and specifications for designing, constructing, inspecting, and operating boilers.

Boiler horsepower. A steaming rate of 3200 lb/h is approximately one boiler horsepower.

Brining. Converting a cation exchange resin to the sodium form by passing a 10 percent solution of sodium chloride through the bed.

British thermal unit. The quantity of heat required to raise the temperature of one pound of water one degree Fahrenheit.

Cation exchange. Replacement of positive ions of salts in solution by sodium ions (softening) or hydrogen ions (demineralizing) by passing through an acidic resin.

Cathode. The negative electrode of an electrochemical cell where electrons are accepted and reduction occurs.

Caustic embrittlement. Cracking of stressed steel in contact with concentrated alkali.

Caustic gouging. Attack on steel by concentrated sodium hydroxide in which hydrogen is evolved and ferrite and hypoferrite ions are formed. Also called ductile gouging.

Chelant. An organic compound capable of forming a cyclic structure in which a metallic ion is bound as a coordination complex.

Chelonate. A coordination complex formed by a chelant and a metallic ion.

Cogeneration. The production of both process steam and electricity in the same plant.

Condensate. Water derived from condensed steam.

Condensing turbine. A turbine from which the driving steam is condensed to water, producing a vacuum.

Contact heater. A device in which water is heated by the direct injection of steam.

Convection. Transmission of heat by propagation through a gas or fluid.

Critical pressure. The pressure at which there is no separation of phase in a fluid—for water, 3203.6 psi.

Deaerator. A heater for removing dissolved oxygen and carbon dioxide by scrubbing feed water with steam.

Decarburization. The oxidation of carbon in the ferrite phase of carbon steel to carbon dioxide.

Demineralization. The removal of both cations and anions from water by ion exchange.

Dew point. The temperature at which a vapor condenses.

Dispersant. A chemical that promotes the suspension of insoluble particles.

Ductile gouging. See caustic gouging.

Economizer. A device for transferring heat from flue gas to feed water to a boiler.

EDTA. Abbreviation for ethylenedinitrilotetraacetic acid.

Enthalpy. The heat, in Btu's, required to raise the temperature of water from 32 F at atmospheric pressure to some other specified temperature and pressure.

Eutectic mixture. A mixture of salts having a minimum melting point less than that of any one of the components.

Exfoliation. The throwing off of thin scales or chips of metal from the surface of an alloy, such as cupro-nickel.

External treatment. Chemical treatments that are accomplished before feed water enters a boiler.

Extraction heater. A heater to which heat is supplied by steam extracted from a turbine.

Extraction turbine. A turbine from which steam is removed at some particular pressure for further use, rather than being condensed.

Feed water. Water fed to a boiler to be converted to steam.

Filming amine. A nitrogen base added to steam to form a film in condensate systems to protect piping from corrosion by oxygen and carbon dioxide.

Flash tank. A vessel into which boiler blowdown water is fed and allowed to vaporize (flash) to steam at some pressure lower than that in the boiler.

Heating value. The number of Btu's per pound of fuel.

Hotwell. Basin or open tank where water from a flash tank is received.

Hydrogen embrittlement. Cracking of steel caused by pressures generated in the reaction between hydrogen and carbon to produce hydrocarbons.

Impulse turbine. A turbine in which the expansion of steam occurs entirely in the nozzles.

Internal treatment. Treatment done inside an operating boiler by injecting chemicals.

Latent heat of vaporization. Heat required at the boiling point to evaporate a pound of water without raising its temperature (970 Btu's).

Magnetite. The protective film of Fe_3O_4 normally present on internal boiler surfaces.

Make up. Treated water required to supplement returned condensate for boiler feed water.

Mud drum. Lower boiler drum where insoluble sludge collects.

Neutralizing amine. Nitrogen base added to steam to neutralize carbon dioxide and raise the pH of condensate.

NTA. Abbreviation for nitrilotriacetic acid.

Over-speed trip. A device that closes the throttle valve on a turbine should its speed exceed a predetermined value.

Oxygen scavenger. A chemical added to reduce the last traces of dissolved oxygen remaining in feed water after deaeration.

Packaged boiler. A shop-assembled steam generator with necessary auxiliary equipment complete and ready to use.

Peaking boiler. A steam generator operated intermittently to meet peak demand for steam and electricity.

Priming. The violent surging of water into the steam outlet caused by sudden ebullition of steam in lower parts of a boiler.

Process steam. Steam used to supply heat for some manufacturing process.

Protective colloid. A chemical that stabilizes colloidal solutions by surrounding or coating colloidal particles, thus preventing their coagulation.

Radiant heat. Heat supplied by radiation without the necessity of some medium for transmission. (e.g., heat from the sun is radiant heat.)

Reaction turbine. A turbine in which a partial reduction in steam pressure takes place in the nozzles, followed by a further reduction in the blades.

Reheat. Returning partially expanded steam to a furnace so that additional heat is added before the steam passes through the final stages of a turbine.

Saponification. The hydrolysis of a fat to a soap by alkali.

Saturated steam. Steam that is saturated with heat at a particular temperature and pressure.

Sludge conditioners. Natural or synthetic protective colloids that prevent the coagulation of particles to form sludge.

Softening. The removal of alkaline earth elements (calcium and magnesium) to prevent scales and sludges from forming within a boiler or feed line.

Spheroidization. The thermal transformation of laminar cementite in the pearlite phase of carbon steel to spherical grains.

Steam drum. The upper boiler drum where feed water and chemicals are injected and blowdown and saturated steam are withdrawn.

Steam purity. Measured by the extent of contamination by solids, either organic or inorganic.

Steam quality. The ratio of the weight of vapor to the total weight of a steam-water mixture.

Steam trap. A device attached to a steam line that automatically drains off any condensate that forms, without releasing any significant amount of steam.

Strainer. A screen to remove sediment that would otherwise plug the orifice in a steam trap.

Supercritical boiler. A boiler operated at a pressure higher than the critical pressure, 3203.6 psi.

Superheat. The number of degrees Fahrenheit of steam above the temperature of saturated steam at some specified pressure.

Superheater. Tubes within a furnace through which saturated steam passes after leaving a boiler to absorb additional heat.

Surface condenser. A water-cooled device for condensing steam.

Surface tension. The resultant attractive forces on molecules at the surface of a liquid, exerted by molecules within the liquid, that tends to make the surface contract to the smallest possible area.

Tube sheet. A supporting plate drilled to hold an array of tubes as, for instance, in a feed water heater or economizer.

Variance. A temporary permit to operate a boiler for six months longer than permitted by the Boiler Code.

Water wall tubes. Tubes containing boiler water that shield refractory furnace walls by absorbing heat.

Appendix 2.

Numerical Problems

1. Disodium phosphate contains 66.9 percent of phosphate. Shortly after adding 5.0 lb of this salt to an operating boiler, a sample of boiler water is taken and found to contain 15 ppm PO_4^{---}. What is the volume of the boiler, in gallons?

   Ans. 26,739 gal

2. After 9 h have elapsed, the boiler water is again sampled and is found to contain 12.4 ppm PO_4^{---}. What is the blowdown rate in gallons per hour?

   Ans. 565 gph

3. If the steaming rate is 75,000 lb/h, what is the feed rate to the above boiler?

   Ans. 9558 gph

4. What is the blowdown ratio?

   Ans. 16.9

5. What is the percentage of blowdown?

   Ans. 5.9 percent

6. Suppose that the average concentration of chloride in a feed water is 10.5 ppm and that in the boiler water is 242 ppm. What is the percentage of blowdown of the boiler?

   Ans. 4.3 percent

7. Supposing the concentration of chloride in the feed water remained constant at 10.5 ppm, what would be its concentration in the boiler water if the blowdown rate were increased to 6.0 percent?

Ans. 175 ppm Cl^-

8. If the blowdown ratio in a certain boiler is 20 and the blowdown rate is 950 gph, what is the steaming rate in pounds per hour?

Ans. 150,537 lb/h

9. What is the feed rate to the boiler in the preceding problem?

Ans. 19,000 gph

10. If the blowdown rate in the same boiler were increased to 8 percent, what would the blowdown rate be in gallons per hour? The feed rate?

Ans. $b = 1569$ gph; $f = 19,619$ gph

11. Suppose one wants to use the lime-soda process to soften water having the following analysis: TH = 125, CaH = 80, MgH = 55, M = 65. What dosage of CaO and Na_2CO_3 should be used?

Ans. 84 ppm CaO; 32 ppm Na_2CO_3

12. How many gallons of water having a total hardness of 88 ppm $CaCO_3$ can be softened by 110 ft^3 of cation exchange resin with an exchange capacity of 30,000 gr $CaCO_3/ft^3$?

Ans. 641,250 gal

13. What will be the cost of softening 1000 gal of this water if NaCl costing \$12.00 per ton is used for regeneration at 14 lb $NaCl/ft^3$?

Ans. \$0.0144/1000 gal

14. It is proposed to use a chelant to dissolve a scale in a boiler. It is known that the solubility product of the scale is $K_{sp} = 2 \times 10^{-11}$ and that the formation constant of the chelonate to be formed is $K_f = 5 \times 10^8$. Will the scale dissolve?

Ans. No

15. A certain raw water has a P alkalinity of 30 ppm and an M alkalinity of 80 ppm $CaCO_3$. What are the approximate concentrations of hydroxyl, carbonate, and bicarbonate ions in this water, in terms of ppm $CaCO_3$?

Ans. $(OH^-) = 0$; $(CO_3^{--}) = 60$; $(HCO_3^-) = 20$

16. A feed water for 900-psi boilers is composed of condensate containing 0.1 ppm SiO_2 and treated make up with 0.9 ppm SiO_2; the combined feed water contains 0.5 ppm SiO_2. What are the percentages of condensate and make up in the feed water?

Ans. 50 percent of each

17. If limited by silica, what is the minimum blowdown that can be used in operating the boilers in Problem 16?

Ans. 2.5 percent

18. Calculate the maximum concentration of silica that can be carried in an 1150-psi boiler to ensure that the concentration in steam does not exceed 0.02 ppm SiO_2.

Ans. 10.7 ppm SiO_2

19. If the saturation temperature of steam from a boiler is 562 F, the P alkalinity is 100, the M alkalinity is 125, and the concentration of silica is 15 ppm, what is the concentration of silica in the steam?

Ans. 0.045 ppm SiO_2

20. If a boiler holds 12,000 gal of water and is being blown down at 220 gph, how long will it take for the concentration of phosphate to decrease from 26 ppm to 13 ppm, assuming that none is added in the meanwhile?

Ans. 37.8 h

Index

Acetylenic alcohols, 227
Acid cleaning, 226-236
   of boilers, 226-236
      preparation for, 228-229
Acid corrosion inhibitors, 226-227
   acetylenic alcohols as, 227
   Mannich bases as, 227
   nitrogen bases as, 227
Acmite, 14, 50, 91, 205, 208
Acridine, 227
Admiralty brass, 140
Air flotation oil separators, 32
Air preheaters, 3, 122, 183
   corrosion in, by sulfuric acid, 199
   inspection of, 196
   regenerative, 183
Alkaline boil-out, 202, 219, 223-224
   analysis of solutions for, 265-266
Alkalinity, 6, 15, 21-22, 40-45, 74-75
   B-reading, 74, 155, 161
      determination of, 245-247
   control of, 21-22, 82-87, 135
      caustic reserve, 22, 75, 81
      congruent, 87
      coordinated phosphate-pH, 23, 66, 71, 75, 81, 82-87, 91, 122, 131
         control limits for, 135
         in marine boilers, 131

   free caustic, 22, 75
   in lime-soda softening, 40-45
   precision, 87
   effect of amines on, 171-172
   effect of ammonia on, 222
   effect of, on silica scaling, 92
   hydroxide, 22
   methyl orange, 15, 39, 74-75, 161-162
      determination of, 245-246
   permissible concentration of, 34
   phenolphthalein, 74-75, 161-162
      determination of, 245-246
      relationships in water, 75
   total, 75
   types of, 74-75

Aluminum, 14, 16
   complexes of, with polyphosphate, 67
Aluminum brass, 132
   corrosion of, 140
Aluminum bronze, 132
Aluminum hydroxide, 16, 31-32, 38-39
   removal of oil by, 31-32
Aluminum oxide, 13
   thermal conductivity of, 13
Aluminum silicate scale, 31
Alundum blasting of turbines, 210

284

Softening capacity, 47-50
Soot blowers, 166, 185, 186
  compressed air for, 185
  effect of, on carryover, 185
  effect of, on steam purity, 166,
    186
  steam for, 185
Spaulding Precipitator, 52-58, 59,
  145, 147-149
  calculation of chemical dosages
    for, 147-149
  composition of sludge in, 57-58
  operation of, 53-58, 59
  rate of flow in, 53-57
  removal of silica in, 57-58, 145
Specific conductance, 167
Spheroidization, 220, 221
Starch, 17, 21, 39, 93
  coagulation of oil by, 30, 93, 202
  coagulation of silica by, 93
Stationary boilers, 86, 120-121
Steam, 4-6, 166-168, 185-187
  contamination in, 12, 34
    prevention of, 23-25
  density of, 121
  determination of sodium in, 166
  efficiency of production of, 182
  extraction, 179
  for generating electricity, 1, 136
  heat content of, 5
  hydrogen in, 168
  isokinetic sampling of, 166
  Larson-Lane purity analyzer for,
    167
  moisture in, 4, 185-187
  particulates in, 166
  properties of, 4-6
  purity of, 4, 166
  quality of, 4, 186-187
  reheating of, 6
  saline contaminants in, 166-168
  saturated, 4-6

  superheated, 4-5, 166, 186
    purpose of, 5
  uses of, 1
  volume of, 6
Steam-blanketing, 168, 219
Steam drums, 3, 15, 18, 20, 31, 35
  characteristics of, 128
  concentration of salts in, 173
  corrosion in, by chelants, 128
  foaming in, 95
  nitrogen-blanketing in, 180
  pitting in, 18, 20, 218
  thinning of, by chelants, 219
Steam ejectors, 29
Steam load, 25, 166, 185
Steam quality, 186-187
  measurement of, by throttling
    calorimeter, 186-187
Stress corrosion cracking, 22-23,
  81-87, 125-126, 221
  factors in, 22-23
  inhibitors of, 81-82, 125-126
  mechanism of, 22-23
  of austenitic alloys, by chloride
    ion, 140, 217
  of turbine rotors, 211
  prevention of, 81-87
Structural homology, 104
Structural iteration, 103
Sulfonated coal, 46
Sulfonated lignins, 17, 21, 23, 93
  charring of, 93
Sulfonated polystyrene, 46, 60
Sulfuric acid, 62
Sulfurous acid, 131
  dissociation of, 131
Sulfur trioxide, 121
Supercritical boilers, 121-122
  operating conditions for, 140-141
Super heat, 4-5, 186

www.ingramcontent.com/pod-product-compliance
Lightning Source LLC
Chambersburg PA
CBHW021030210326
41598CB00016B/976